Wei Chen

Fault detection and isolation in nonlinear systems

Wei Chen

Fault detection and isolation in nonlinear systems

Wei Chen

Fault detection and isolation in nonlinear systems

Observer and energy-balance based approaches

Südwestdeutscher Verlag für Hochschulschriften

Impressum/Imprint (nur für Deutschland/only for Germany)
Bibliografische Information der Deutschen Nationalbibliothek: Die Deutsche Nationalbibliothek verzeichnet diese Publikation in der Deutschen Nationalbibliografie; detaillierte bibliografische Daten sind im Internet über http://dnb.d-nb.de abrufbar.
Alle in diesem Buch genannten Marken und Produktnamen unterliegen warenzeichen-, marken- oder patentrechtlichem Schutz bzw. sind Warenzeichen oder eingetragene Warenzeichen der jeweiligen Inhaber. Die Wiedergabe von Marken, Produktnamen, Gebrauchsnamen, Handelsnamen, Warenbezeichnungen u.s.w. in diesem Werk berechtigt auch ohne besondere Kennzeichnung nicht zu der Annahme, dass solche Namen im Sinne der Warenzeichen- und Markenschutzgesetzgebung als frei zu betrachten wären und daher von jedermann benutzt werden dürften.

Coverbild: www.ingimage.com

Verlag: Südwestdeutscher Verlag für Hochschulschriften GmbH & Co. KG
Heinrich-Böcking-Str. 6-8, 66121 Saarbrücken, Deutschland
Telefon +49 681 37 20 271-1, Telefax +49 681 37 20 271-0
Email: info@svh-verlag.de

Approved by: Duisburg, Duisburg-Essen, Uni, Diss., 2011

Herstellung in Deutschland (siehe letzte Seite)
ISBN: 978-3-8381-3327-0

Imprint (only for USA, GB)
Bibliographic information published by the Deutsche Nationalbibliothek: The Deutsche Nationalbibliothek lists this publication in the Deutsche Nationalbibliografie; detailed bibliographic data are available in the Internet at http://dnb.d-nb.de.
Any brand names and product names mentioned in this book are subject to trademark, brand or patent protection and are trademarks or registered trademarks of their respective holders. The use of brand names, product names, common names, trade names, product descriptions etc. even without a particular marking in this works is in no way to be construed to mean that such names may be regarded as unrestricted in respect of trademark and brand protection legislation and could thus be used by anyone.

Cover image: www.ingimage.com

Publisher: Südwestdeutscher Verlag für Hochschulschriften GmbH & Co. KG
Heinrich-Böcking-Str. 6-8, 66121 Saarbrücken, Germany
Phone +49 681 37 20 271-1, Fax +49 681 37 20 271-0
Email: info@svh-verlag.de

Printed in the U.S.A.
Printed in the U.K. by (see last page)
ISBN: 978-3-8381-3327-0

Copyright © 2012 by the author and Südwestdeutscher Verlag für Hochschulschriften GmbH & Co. KG and licensors
All rights reserved. Saarbrücken 2012

Acknowledgments

This work is carried out during my stay at the Institute of Automatic Control and Complex Systems (AKS), University of Duisburg-Essen for acquiring my Ph.D degree. It is a great pleasure for me to thank those who have given me valuable supports and encouragements during the last years.

First and foremost, I would like to express my deepest gratitude to Prof. S. X. Ding. His perpetual encouragement, guidance and support enabled me to complete this work. I am also greatly indebted to Prof. Vincent Cocquempot for his interest in my work. His valuable suggestions and constructive remarks added to the quality of this thesis.

I would like to thank my groupmates Dr. Abdul Qayyum Khan and Dr. Muhammad Abid for their support and long hour discussions. Many thanks to all of my colleagues at AKS for creating an inspiring and pleasant atmosphere. Special thanks to Dipl.-Ing. Georg Nau, Dipl.-Ing. Christoph Kandler and M.Sc. Ali Abdo for the valuable discussions and support for my work.

Finally, I would like to express my gratitude to my parents for their understanding and support for my long years study and living far away from them.

Duisburg, November 2011
Wei Chen

Acknowledgments

To my parents

Contents

Acknowledgments	i
Abbreviations and notations	vi
Abstract	viii

1 Introduction   1
    1.1 Motivation 1
    1.2 Objectives of the thesis 3
    1.3 Outline and contribution of the thesis 4

2 State of the art of fault diagnosis techniques   6
    2.1 Some basic concepts 6
    2.2 Classification of fault diagnosis techniques 7
       2.2.1 Hardware redundancy based fault diagnosis 7
       2.2.2 Signal processing based fault diagnosis 7
       2.2.3 Plausibility test 8
       2.2.4 Model-based fault diagnosis 8
       2.2.5 Summary of fault diagnosis techniques 11
    2.3 Model-based FDI schemes for nonlinear systems 11
       2.3.1 Observer-based approaches 12
       2.3.2 Approaches based on energy-balance and passivity 13
    2.4 Summary 14

3 Observer-based FD for Lipschitz nonlinear systems: an integrated desgin approach   15
    3.1 Introduction 15
    3.2 Preliminaries 16
    3.3 Definitions of FAR and FDR 20
       3.3.1 Review of the definitions in the statistical framework 20

		3.3.2 Definitions in the norm-based framework	21

- 3.4 Integrated design of FD systems 21
    - 3.4.1 Problem formulation 21
    - 3.4.2 A solution to the integrated design problem 23
- 3.5 A design example 30
    - 3.5.1 System description 30
    - 3.5.2 Simulation results 31
- 3.6 Summary 34

4 Energy-balance based FDI framework for passive nonlinear systems 35
- 4.1 Introduction 35
- 4.2 Preliminaries 37
- 4.3 Energy balance of passive systems 37
- 4.4 Definition of faults 39
    - 4.4.1 Stored-energy change 39
    - 4.4.2 Dissipated-energy change 40
- 4.5 Fault detection 41
- 4.6 Fault isolation 41
    - 4.6.1 Isolation of different energy changes 42
    - 4.6.2 Isolation of different components 43
- 4.7 Energy-balance based FDI for passive systems with unknown inputs 44
    - 4.7.1 Fault detection 45
    - 4.7.2 Fault isolation 45
- 4.8 RLC Example 47
    - 4.8.1 System description 47
    - 4.8.2 Energy balance construction 48
    - 4.8.3 Fault detection 49
    - 4.8.4 Fault isolation 50
    - 4.8.5 Parameter setting and threshold computation 51
    - 4.8.6 Simulation results 52
- 4.9 Summary 57

5 Design of Energy-balance based FDI for two classes of passive nonlinear systems 58
- 5.1 Input-affine passive systems 58
    - 5.1.1 Introduction 58

		5.1.2	Energy balance of input-affine passive systems	59
		5.1.3	A design example	60
		5.1.4	Application to linear passive systems	62
		5.1.5	Input-affine passive systems with unknown inputs	65
	5.2	Lagrangian systems		67
		5.2.1	Introduction	67
		5.2.2	Lagrangian systems with external forces and dissipation	68
		5.2.3	Stored energy of Lagrangian systems	68
		5.2.4	Passivity of Lagrangian systems	69
		5.2.5	Energy balance of Lagrangian systems	70
		5.2.6	Lagrangian systems with unknown inputs	70
	5.3	Summary		72

6 Application to the robot manipulator benchmark — 73

	6.1	Introduction	73
	6.2	System model and description of faults	74
	6.3	Energy-based FDI system design	76
		6.3.1 Energy balances construction	76
		6.3.2 Fault detection	79
		6.3.3 Fault isolation	80
	6.4	Simulation	84
		6.4.1 Parameter setting and thresholds computation	84
		6.4.2 Simulation results	84
	6.5	Summary	86

7 Conclusions and future directions — 97

	7.1	Conclusions	97
	7.2	Future directions	98

Bibliography — **100**

Abbreviations and notations

Abbreviations

Abbreviation	Meaning
FAR	false alarm rate
FD	fault detection
FDI	fault detection and isolation
FDIA	fault detection, isolation and analysis
FDR	fault detection rate
LMI	linear matrix inequality
NMI	nonlinear matrix inequality
SDF	set of detectable fault
SDFA	set of disturbances that cause false alarms

Notations

Symbol	Description
$\forall$	for all
$\in$	belong to
$\implies$	implies
$\iff$	equivalent to
inf	infimum
min	minimum
$\Re^n$	space of real n-dimensional vectors
$\Re^{n \times m}$	space of n by m matrices
$\lvert . \rvert$	absolute value of a scaler
$\lVert . \rVert$	Euclidean norm of a vector
$\lVert . \rVert_2$	$\mathcal{L}_2$ norm of a (vector-valued) signal
$\lVert . \rVert_-$	H_- index of a system

Σ	a dynamic system
Δ	model uncertainty
$\mathscr{E}$	system energy
$\mathscr{H}$	energy balance of a passive system
Λ	a diagonal matrix
a^T	transpose of a vector a
A^T	transpose of a matrix A
$A > 0$ ($A \geq 0$)	the symmetric matrix A is positive (semi) definite
$A < 0$ ($A \leq 0$)	the symmetric matrix A is negative (semi) definite
$prob(a < b)$	probability that $a < b$
I	identity matrix
L_a	Lagrangian function of a system
J	evaluated residual signal
J_{th}	threshold
d	unknown input vector (disturbance)
f	fault vector
r	vector of residual signals
u	input vector
x	state vector
$\hat{x}$	estimation of the state vector x
y	output vector

Abstract

Due to the increasing complexity of modern technical processes, the most critical issues in the design of an automated system nowadays are safety/reliability, higher performance and cost efficiency. Faults in process components can lead to a considerable reduce of the efficiency of the process, quality of the product and in some cases even result in fatalities. In order to avert these losses, an efficient diagnosis of the faults plays a central role. Therefore, fault diagnosis is becoming an essential part of modern control systems. Fault diagnosis of linear dynamical systems has been extensively studied since decades and well-established techniques exist in the literature. However, fault diagnosis for nonlinear dynamical systems is yet an active field of research. Since most of real systems are nonlinear in nature, classically, linear fault diagnosis techniques have been applied to nonlinear systems based on the linearized system model around an operating point. The drawback of this approach is the limited fault diagnosis performance. In order to fulfill the increasing demand of more effective fault diagnosis systems for nonlinear processes, a lot of attention has been paid to nonlinear fault diagnosis techniques, which is the major topic of this thesis.

Different from linear systems, there is no uniform solution for the fault diagnosis of general nonlinear systems. Various schemes have been proposed for nonlinear systems with special structures. Among them, Lipschitz nonlinear systems have been intensively studied, since on one hand more general nonlinear systems can be transformed into Lipschitz nonlinear systems, and on the other hand, many linear fault diagnosis approaches can be extended to these kind of nonlinear systems. For Lipschitz nonlinear systems, observer-based fault detection approach has been mostly applied, which consists of an observer-based residual generator and a residual evaluator. Classically, residual generator and residual evaluator are designed separately. Since the performance of fault detection system is decided by residual generator and evaluator together, it can be expected that, higher fault detection performance can be achieved by designing these two units in an integrated manner instead of separate handling of them. Motivated by this fact, an integrated design approach of observer-based residual generator and evaluator is proposed for Lipschitz nonlinear systems.

Besides the schemes extended from linear methods (i.e. observer-based approach, parity space approach etc.), new nonlinear fault diagnosis techniques have also been studied recently, which can be effectively applied to complex nonlinear systems i.e. switched nonlinear systems, hybrid nonlinear systems etc. Among them, new fault diagnosis schemes based on passivity and energy-balance which are closely related to system "energy" have a great potential due to their clear physical meanings. In this thesis, this approach is extended to a complete fault detection and isolation framework with the focus on passive nonlinear systems.

The fault diagnosis methodologies proposed in this thesis are tested with the desgin examples in the respective chapters and with the robot manipulator benchmark problem. The simulation results show the effectiveness of the proposed schemes.

Chapter 1

Introduction

This chapter briefly describes the motivations and objectives of this thesis. The outline and contribution of the thesis are also presented.

1.1 Motivation

Fault detection and isolation. Modern technical processes are becoming more and more complex. Consequently, the demand for safety/reliability, higher performance and cost efficiency are of major importance in the design of an automated system nowadays. Faults can occur in all kinds of process components, for example, pressure drop out in hydraulic components, leakages in pipes, drifting of sensors etc. Faults may lead to a considerable reduce of the efficiency of the process, quality of the product and in some cases even result in complete failure of the process. In [1], it is claimed that 20 billion dollars per year is lost due to poor abnormal situation management only in US petrochemical industry. For safety-critical systems such as aircrafts, nuclear reactors etc., faults may lead to catastrophic incidents which result in big loss of human lives, a few incidents are listed below:

> Boeing 747-200F lost both engines on taking off from Schiphol Airport in Amsterdam. After 15 minutes, the crew lost the control and the plane crashed into a building with a considerable loss of life.

> The American Airline DC10 crashed at Chigao-OHare International Airport. The pilot had the indication of fault only 15 seconds prior to the accident.

> An explosion happened in a huge nuclear power plant in the town of Chernobyl in 1986.

Surprisingly, it has been shown in investigations and research reports [2–5] that, these incidents could have been avoided if there was a suitable fault monitoring and tolerant system. For

Chapter 1 Introduction

example, Maciejowski [2] has shown that the first incident of Boeing 747-200F could have been avoided by proper reconfiguration of the controller. Motivated by the above mentioned reasons, the development of fault detection and isolation (FDI) is receiving more and more attentions both in academia and industry, for instance, [6–11] to name a few.

Model-based FDI. The so-called *redundancy* plays an essential role for a successful FDI. System redundancy can be realized by reconstruction of the process components using the identical (redundant) hardware components. Faults are then detected by the deviation between the actual process output and the output of redundant process component. Since the use of redundant hardware results in high costs, its application is only restricted to a number of key components [6]. Another kind of system redundancy is called *analytical redundancy*, which is constructed based on a system model. The model which represents the system properties is built based on the knowledge of the system. The fault detection is achieved by comparing the behavior of the system with its model. Since analytical redundancy can be realized in the same processor which implements the control algorithms, thus no additional hardware is needed. Due to associated advantages, analytical redundancy based approaches are becoming more and more popular, and its efficiency has been demonstrated by a great number of successful applications in industrial processes and automatic control systems, for instance, vehicle control systems, robots, power systems etc.

Nonlinear systems. Model-based FDI approaches for linear systems have been well studied since decades, and quite a large number of methods exist in literature. Since most pratical systems are nonlinear in nature, the linear FDI schemes are applied usually based on the linearization of the nonlinear systems at different operating points, and the linearization errors are modeled as unstructured uncertainties. The disadvantages of this approach are: (1) It is difficult or even impossible to prove the stability of the FDI system, since it is based on an approximated system model; (2) Because only part of the informations of nonlinear system model (linearized model) is used, the performance of FDI schemes is limited. Therefore, FDI techniques for nonlinear systems have received a lot of attentions and is a quite active field of research recently, see for instance [12–14] and the references therein.

Generally speaking, there are three types of model-based FDI approaches including the observer-based approach, parity-space approach and the parameter identification approach. In the past few decades, observer-based methods have received considerable attention, due to their advantages of early detection, easy on-line implementation etc [6, 7]. For nonlinear systems, because of the difficulties in stability/convergence analysis, the application of the classical model-based FDI schemes faces a lot of challenges. Since an applicable uniform solution for general nonlinear systems is still an open issue, nonlinear systems with special structure are intensively studied. One kind of widely used nonlinear systems is systems with Lipschitz nonlinearities, since on one hand more general nonlinear systems can be transformed into Lipschitz nonlinear systems under some conditions [15], and on the other hand, many linear FDI approaches can be extended to this kind of nonlinear systems [16–18]. Recently, advanced nonlinear FDI methods have been developed, among them, passivity and energy-balance based approaches have a great potential due to its clear physical meaning and easy implementation [19–21]. These approaches are closely related to system "energy", where the "energy" could be true physical energy or an abstract energy function (Lyapunov function) defined in Lyapunov theory [22].

1.2 Objectives of the thesis

In this thesis, new nonlinear FDI schemes based on observer and energy-balance are studied. The first intention is to develop a more efficient design approach of observer-based FD for Lipschitz nonlinear systems. The second intention is to establish a complete energy-balance based FDI framework for passive nonlinear systems. In the following, the concrete objectives are presented.

Observer-based FD for Lipschitz nonlinear systems. The idea of observer-based fault detection is to generate estimations of measured system outputs based on system model, and then the estimations are compared with the real measurments to generate a so-called residual signal, which carries the information of faults. Ideally, when there are no modeling error and disturbances, the estimations will converge to the measurements in fault-free case and the residual signal will go to zero. After the transient, a non-zero residual signal indicates the occurance of faults. However, since modelling error and disturbances are inevitable, a threshold should be designed to distinguish the non-zero residual signal driven by faults from that caused by modelling error/disturbances. In practice, essential requirements on a fault diagnosis system are generally expressed in terms of a lower *false alarm rate* (FAR) and a higher *fault detection rate* (FDR), and an optimal trade-off between them is of primary interest in designing an FD system [6]. In this context, the traditional way of separately designing residual generation, evaluation and threshold computation makes less sense. To achieve a successful design of a observer-based FD system, an integrated handling of residual generator, evaluator and threshold is needed. The integrated design of FD for linear systems has been well studied in [23, 24], the first objective of this dissertation is to extend the integrated design approach to Lipschitz nonlinear systems, which is summarized as: proposing an integrated design approach for fault detection of Lipschitz nonlinear systems which achieves an optimal trade-off between low FAR and high FDR.

Energy-balance based FDI for passive nonlinear systems. Passivity-based fault detection and fault tolerance analysis have been newly introduced by [21] for switched nonlinear systems, where the passivity is represented by an energy relation including the stored energy in the system and the supplied energy from the outside. The fault detection is then carried out by checking this passive energy relation. In [19, 20], energy balance of steady-state processes is used for fault detection system design. Motivated by these studies, the second intention of this thesis is to establish a new energy-balance based FDI framwork for passive nonlinear systems. Compared with the energy relation used in [21], an energy balance which also includes the dissipated energy is constructed. Based on it, a better fault detectability can be achieved. In order to make full use of the energy properties of passive systems, faults should be defined according to their different influences on system energies, and new isolation schemes based on the "energy properties" of faults need to be developed. These objectives can be summarized in the following points:

Establishing the energy balance of passive nonlinear systems.

Definition of faults according to their different influences on system energies.

Proposing energy-balance based fault detection and isolation schemes.

1.3 Outline and contribution of the thesis

In the following, the structure and major contributions of the thesis are briefly outlined.

Chapter 2 Background and state of the art presents a review of fault diagnosis techniques. Fundamental terminology, such as faults, unknown inputs, fault detection, fault isolation, and fault identification are introduced according to the recommendation of IFAC technical committee SAFEPROCESS [25]. A widely accepted classification of fault diagnosis techniques and a comparision between them are presented. A comprehensive survey on observer-based and energy-balance based fault diagnosis techniques for linear and nonlinear systems is given which is the focus of this dissertation.

Chapter 3 Observer-based FD for Lipschitz nonlinear systems: an integrated desgin approach proposes an integrated design approach of observer-based FD for Lipschitz nonlinear systems, which results in a trade-off between low FAR and high FDR. The extension of concepts FAR and FDR to the norm-based framework is introduced, which includes *set of disturbances that cause false alarms, set of detectable faults* and the simplified norm-based definition of FDR [6]. Sufficient conditions for the existence of FD system which achieves the given FDR and FAR are derived in the form of linear matrix inequalities. Based on it, an iterative algorithm is presented which minimizes the FAR for a given FDR. Finally, the proposed approach is illustrated by a design example.

Chapter 4 Energy-balance based FDI framework for passive nonlinear systems establishs an energy-balance based FDI framework for passive nonlinear systems. Based on passivity, an energy balance of passive nonlinear systems is constructed, which includes stored, dissipated and supplied energies. In the new FDI framework, faults are defined according to their different influences on system energies. The fault detection is carried out by checking the validity of the energy balance. The proposed fault isolation schemes include two steps: (1) Isolation of different classes of faults (energy changes); (2) Isolation of different components. For passive nonlinear systems with unknown inputs, thresholds for fault detection and isolation are designed based on the bound of "unknown-input energy". An RLC circuit example is given to illustrate the proposed FDI framework.

Chapter 5 Design of energy-balance based FDI for two classes of passive nonlinear systems proposes design procedures of the energy-balance based FDI schemes for two classes of nonlinear systems: (1) Input-affine passive systems; (2) Lagrangian systems. The design procedures include finding a storage function, establishing the energy balance and computation of the threshold. The proposed design procedure is also applied to linear passive systems, which can be considered as a special case of input-affine passive systems.

Chapter 6 Application to the Robot Manipulator benchmark presents the application of the energy-balance based FDI schemes to robot manipulator benchmark. This benchmark has strong nonlinearities and is widely used for factory automation systems. Since in practice robot manipulators are usually driven by DC motors, the dynamics of DC motors is also considered in the benchmark model. This benchmark can be modeled as an interconnection of two Lagrangian systems, which is an excellent benchmark for illustrating the usefulness of the proposed energy-balance based FDI framework.

Chapter 2

State of the art of fault diagnosis techniques

> *This chapter reviews fault diagnosis techniques. Fundamental terminology, such as faults, unknown inputs, fault detection, fault isolation, and fault identification are introduced. A widely accepted classification of fault diagnosis techniques is presented. Model-based fault diagnosis for nonlinear systems is the focus of this thesis, therefore, a particular attention has been given to the state of the art of model-based fault diagnosis techniques for nonlinear systems.*

2.1 Some basic concepts

The terminology used in this thesis is fairly standard based on the recommendation of IFAC technical committee SAFEPROCESS. In the following, basic definitions of faults, uncertainties, disturbances, and the descriptions of fault detection, fault isolation and fault analysis/identification are given. The detailed explanation of the above mentioned terminology can be found in [6, 25].

A *fault* is an un-permitted deviation of at least one characteristic property or parameter of the system from the standard/acceptable condition. It can be modeled as an external input or as parameter deviation which changes the system characteristics. Mathmatically, *uncertainties* and *disturbances* can be modeled quite similar to faults as parameter deviation and external input. The essential difference between them is that, uncertainties and disturbances are unavoidable and are present during the normal operation of system, so they should be taken into account in the control system design. Faults, on the other hand, are considerded as a special situation in the system design since they appear at an unknown time point and usually lead to severe changes in the system. It is either impossible or too conservative to consider all the faults in the design of controller which also works for normal system operation, so a detection of faults is needed in order to change the control system from the normal operation to a faulty-mode. In this special mode, proper change of the control system should be made to reduce the effects of faults on the system.

Generally speaking, fault diagnosis consists of the following three essential tasks [6]:

Fault detection: detection of the occurrence of faults.

Fault isolation: localization (classification) of different faults.

Fault analysis or identification: determination of the type, magnitude and cause of the fault.

Depending on the function, a fault diagnosis system is called FD or FDI or FDIA (fault detection, isolation and analysis) system [6]. Since the existence conditions for fault identification are very strict, FD and FDI systems have been more intensive studied, which is also the focus in this thesis.

2.2 Classification of fault diagnosis techniques

A number of fault diagnosis techniques have been developed during the last decades. In sequel, a rough classification of these techniques is presented.

2.2.1 Hardware redundancy based fault diagnosis

The essential idea of this scheme is to reconstruct the process components using the redundant hardware components. Since when there is fault in the process component, the output of the process component will be different from the one of its redundancy, the fault detection is achieved by compare the outputs from the process component and its redundancy. The high reliability and ability of direct fault isolation are the main advantages of this scheme. However, the use of redundant hardware leads to high costs. Thus, its use is limited to a number of key components and especially for the applications which have a very high safety requirement, for example, aerospace task, flight control systems, etc. [6, 7].

2.2.2 Signal processing based fault diagnosis

The core of this scheme is to extract the information of the faults from the process signals. Assume that certain process signals carry information about the faults of interest in form of symptoms, fault diagnosis can be achieved by a suitable signal processing. Typical symptoms can be time domain functions or frequency domain functions. Time domain symptoms are for example magnitude, mean values (arithmetic or quadratic), limit values, trends, and statistical moments of the amplitude distribution or envelope, etc., while the frequency domain symptoms include spectral power densities, frequency spectral lines and ceptrum to list a few. Since the dynamics of the process have been not taken into account in the Signal processing based fault diagnosis, this approach is mainly used for processes in the steady state [6].

2.2.3 Plausibility test

The basic idea of this technique is to evaluate the measured process variable with regard to credible, convincing values or their compatibility among each other based on some simple physical laws. On the assumption that a fault leads to the loss of plausibility, the information about the presence of a fault can be extracted by the plausibility check. This approach can be viewed as a first step to model-based fault detection since the physical laws represent a part of the model information. However, the efficiency of this approach is limited for detecting faults in a complex process and for fault isolation [6, 10].

2.2.4 Model-based fault diagnosis

The intuitive idea of the model-based FD technique is to replace the hardware redundancy by a process model which is implemented in the form of software. The process model can be an analytical model represented by set of differential equations or it can be knowledge-based model represented by, for example, neural networks, petri nets, experts systems, fuzzy rules etc [26]. With process model, the process behavior can be reconstructed on-line. Analogous to hardware redundancy, it is called software redundancy or analytical redundancy.

In the framework of the model-based fault diagnosis, the process model runs in parallel to process and is driven by the same process inputs. The information about the fault is obtained by comparing the measured process variables (output signals) with their estimates delivered by the process model. The difference between the measured process variables and their estimates is called *residual*. When we have a exact process model, the residual should be zero in the fault-free operating states and become non-zero by a fault in the process. In this case, fault detection can be carried out by checking the residual. The procedure of creating the residual signal is called *residual generation* . Correspondingly, the process model and the comparision unit build the so-called *residual generator*. Residual generation can be considered as an extended plausibility test, where the plausibility is the complete process input-output behavior which is represented by process model. The plausibility check here is the comparision of the process outputs with their estimates [6].

Since modeling error and unknown disturbances are inevitable for technical process, the generated residual signal is usually non-zero even in the fault-free case. So a post-processing of the residuals which extracts the information about the fault of interests from the residual signals is needed, which is called *residual evaluation*. Residual generation and residual evaluation builds the core of the model-based fault diagnosis technique. In the following, different residual generation and evaluation approaches are introduced.

Observer-based residual generation

In observer-based residual generation approaches, the estimation of the process output is generated by an observer. It should be noted that the design of observers used for FD purposes is different from that for control purposes. The observers needed for control are state observers which is used to estimate the unmeasured states. In contrary, the observers needed for FD (diagnostic observers) are output observers which estiamte the measured states. Generally, the existence conditions for diagnostic observers are much more relaxed than that for a state observer. Full state observers like fault detection filter are also widely used for residual generation, the extra design freedom is used to achieve fault isolation, unknown input deoupling etc.

In order to achieve an optimal residual generation, considerable effort has been devoted to develop observer-based residual generator which fulfills the following two requirements: (1) robust to model uncertainties, disturbance and sensor noises; (2) sensitive to faults. Unknown input observers which decouple the residual signal from the unknown disturbances were introduced by the pioneering work of Wnnenberg and Frank in [27] and then considerable contribution was made in [28–30]. The drawback of this approach is the hard existing conditions and the reduce of the fault detectability. In order to make a better trade-off between robustness and sensitiveness, instead of completely decoupling unknown inputs, much focus has also been paid to design observers which attenuate the effect of unknown inputs based on H_2 or H_∞ index [31, 32]. Multi-objective optimization approach which consider the robustness and sensitivity problem simuteniously are studied in [33–35] based on the index H_-/H_∞, where H_- is the measurement of the fault sensitivity and H_∞ index represents the robustness. Recently, [36, 37] have proposed an unified solution which solves the H_i/H_∞ (including H_-/H_∞ and H_∞/H_∞) optimization problem, where H_i represents all nonzero singular values of the transfer matrix from faults to residual signal.

Parity space based residual generation

In parity space based residual generation approaches, a so-called parity relation is derived from system model. The residual generation is achieved by checking the consistency of the parity relation based on a so-called parity vector. This approach has been first proposed by [38] for state space model of the system, later contributions were made using the transfer functions in [39–41].

There is a close relationship between parity space approach and observer-based approach. As mentioned in [6], parity space approach leads to certain types of observer structures and is therefore structurally equivalent to the observer-based approach, even though the design procedures are different.

Parameter identification based methods

The core of the parameter identification based methods is an on-line parameter estimation. The fault decision is performed by comparing the estimated parameter with the nominal process parameter. An advantage of the approach is that it yields the size of the deviations of process parameters which is very useful for fault analysis [26]. The disadvantage is that, in order to estimate the parameters correctly a sufficient excitation is needed, which is not always available [26]. There are a number of parameter identification based schemes, for example, methods based on least squares (LS), recursive least squares (RLS), extended least squares (ELS), etc.

Residual evaluation schemes

Based on the type of system under consideration, the evaluation schemes can be roughly divided to statistical-based methods and norm-based methods. For stochastic systems, the stochastical properties like mean, variance, likelihood ratio (LR), generalized likelihood ratio (GLR) are used for the evaluation of residuals [42–44]. For deterministic systems, the norm-based residual evaluation is employed, where different kinds of norm like $\mathscr{L}_2$, peak and also Root Mean Square value (RMS) are used [6]. Besides requiring less on-line computation, norm-based schemes also allow a systematic way for threshold computation.

Integrated design of residual generation and residual evaluation

In most of the studies, these two parts have been designed separately. Since the performance of fault detection system is decided by residual generator and evaluator together, a residual generator optimized under some performance index does not automatically result in an optimal fault detection system. So in order to achieve an optimal FD performance, it is necessary to study the approach which design the residual generator and evaluator in an integrated manner instead of separate handling of these units. [23, 24] have proposed the integrated design appraoch of FD system for linear systems, which achieves an optimal trade-off between a lower *false alarm rate* and a higher *fault detection rate*.

2.2.5 Summary of fault diagnosis techniques

The select of a proper FDI scheme depends on many factors. First of all, the properties of process dynamics should be considered. For example, signal processing based fault diagnosis can be only used for processes working in the steady-state or with slow dynamics. Secondly, the availability of a process model is also an important factor. For example, it is relative easy to obtain a mathmatical model of mechanical systems, where analytical model-based FDI schemes can be applied. However, for processes like in chemical industry it is very hard or even impossible to get an analytical model. In this case, qualitative model-based approaches or signal-based approaches should be applied. Thirdly, the choice of FDI schemes is also decided by the requirements like detection time, ability to isolate the faults, etc. Approaches like Plausibility test can only achieve the fault detection. Generally speaking, when a process model is available and the requirement of FDI performance is relative high, then the model-based FDI schemes are the best choice since it has the advantage of low cost, fast detection, ability to deal with fast dynamical systems and also able to isolate or even identify the faults. So that, the focus of this dissertation is model-based FDI schemes, specially for nonlinear systems. The state of art of this area is reviewed in the next section.

2.3 Model-based FDI schemes for nonlinear systems

Most of the real systems are nonlinear in nature. One widely applied approach for the fault detection of nonlinear systems is to linearize the nonlinear model at a number of operating points, and for each operating point, the well established theory of linear model-based FDI is used. In this framework, the linearization error is considered as model uncertianties and handeled by applying robust FDI schemes. The drawback of this approach is firstly the FDI performance, since the model uncertianties could be very large due to the linearization, the threshlod should be set very high in order to avoid false alarms, which will on the other hand reduce the fault detectability. Secondly, since FDI systems designed for different operating point are switched between each other and also because of the linearized models, the stability of the whole FDI system is very hard or even impossible to guaranty. These limitations of applying linear FDI methods to nonlinear systems motivated the researchers to study the nonlinear FDI techniques.

The model-based nonlinear FDI schemes are mostly extended from linear FDI theory, which includes observer-based approaches, parity space approaches and parameter identification based approaches. Among them, observer-based approach has been most intensively studied and is also one of the focuses in this thesis, so the important methods of this approach will be reviewed. Recently, new nonlinear FD schemes have been introduced in [19, 20] based on energy conservation of nonlinear industrial processes and in [21] using the passive properties of nonlinear systems. These approaches are closely related to system energy. Due to their clear physical meaning and easy implementation, they have a great potential for the nonlinear FDI system design. The pioneer studies in this field will also be introduced.

2.3.1 Observer-based approaches

High-gain observer approach

Proposed in [45–48], high-gain observer approach is developed for the input affine nonlinear systems based on a nonlinear transformation described in [46]. Based on the transformed system model, a nonlinear observer can be designed whose observer gain is obtained by solving a linear algebraic equation. This approach can be applied to a large class of nonlinear systems and the observer desgin is carried out in a systematic way. However, the drawback is the high sensitivity of the nonlinear transformation to model uncertainties, and also the peaking phenomenon because of the normally very high observer gain.

Sliding mode observer approach

The inherent property of sliding mode observers is its robustness to uncertainties and disturbances [49–51]. It can be applied to nonlinear systems whose dynamics include a linear part and a nonlinear part which is Lipschitz with respect to system states, the uncertainties and disturbances are assumed to be bounded by an known Lipschitz nonlinear function. The design of sliding mode observer consists of two steps: construction of a sliding surface and designing a feedback law which drives the system trajectories to the sliding surface in finite time. As the trajectories reach to the sliding surface, the estimation becomes insensitive to the external disturbances. The limitation of this approach is its requirement of sufficient measurements and the chattering phenomenon caused by the nonlinear feedback in the observer.

Geometric approach

Extended from the detection filter proposed by Massoumnia [52], the nonlinear geometric approach for fault detection is studied in [53–55]. The basic idea is to construct a subsystem which is decoupled from disturbances and only affected by faults. This is achieved by finding an unobservability subspace in which all disturbances are unobservable. Observer is then designed for the subsystem, where the disturbances are decoupled. For fault isolation, all the other faults, except the one to be isolated, are treated as disturbances and are rendered unobservable to the subsystem. A recursive algorithm is proposed in [53] to find the unobservability subspace. The drawback of this approach is that, the existing conditions are very hard to fulfill, and the fault detectability is reduced when part of faults belong to the same subspace as the disturbances.

Observers for Lipschitz Nonlinear systems

The formulation of Lipschitz nonlinear systems has been widely used in the development of nonlinear FD techniques, since under some conditions, more general nonlinear systems can be transformed into Lipschitz nonlinear system as discussed in [15]. The dynamics of Lipschitz nonlinear systems contain a linear part and a nonlinear part which fulfills the Lipschitz condition. Based on the well established linear matrix inequality (LMI) techniques, the optimal residual generation problem can be solved for Lipschitz nonlinear systems, which achieves the trade-off between robustness against disturbances and sensitiveness to faults [16–18, 56–59]. The first objective of this dissertation is to extend the integrated FD design approach, which is well developed for linear systems, to Lipschitz nonlinear systems.

2.3.2 Approaches based on energy-balance and passivity

In [19, 20], energy-balance based approaches have been introduced for the FDI system design of complex industrial processes, where an input-output model of the system is impossible to obtain. Assume the process works in the steady-state or around an operating point, the energy conservation principle are used to establish an energy balance of the process, which can be considered as an "energy model". Based on it, parameter estimation approaches and parity space approaches can be used to generate the residual signal. In [21], a FD scheme which connects the energy and the passivity has been proposed for nonlinear switched systems. The concept passivity relates closely to system energies, its physical meaning is that, the energy stored in passive systems can not be more than the energy supplied by the environment outside ([21]). This property can be expressed by the following inequality:

$$V(x(\tau)) - V(x(0)) \leq \int_0^\tau y^T u\, dt \qquad (2.1)$$

where nonnegative function V with $V(0) = 0$ is called storage function which represents the energy stored in the system, and $y^T u$ is called supplied rate which represents the energy supplied from the outside. x are system states. (2.1) is always fulfilled for passive systems in fault-free case, so fault detection can be carried out by checking (2.1), when it fails, it means there are faults in the system.

2.4 Summary

In this chapter, a review of fault diagnosis techniques was given. Definitions of basic terminology such as fault, unknown inputs, fault detection, fault isolation, and fault identification were presented. The main ideas of different kinds of fault diagnosis techniques including their merits and demerits were introduced. A particular attention has been given to model-based approaches. A detailed survey of observer and energy based approaches has been presented, which is the focus of this thesis.

Chapter 3

Observer-based FD for Lipschitz nonlinear systems: an integrated desgin approach

> The objective of this chapter is to propose an integrated design approach of observer-based FD for Lipschitz nonlinear systems, which results in a trade-off between low FAR and high FDR in the norm-based framework. The original definitions of FAR and FDR in the statistic framework are first reviewed, then the extension of these two concepts to the norm-based framework is introduced, which includes set of disturbances that cause false alarms ($SDFA$), set of detectable faults (SDF) and the simplified norm-based definition of FDR [6]. The FAR in the norm-based framework is introduced in the way that, $SDFA$ can be minimized. Sufficient conditions for the existence of FD system which achieves the given FDR and FAR are derived in the form of linear matrix inequalities. Based on it, an iterative algorithm is presented, which minimizes the FAR for a given FDR. Finally, a design example is given to illustrate the proposed approach.

3.1 Introduction

Among model-based FD schemes, observer-based approaches have been most intensively studied, see for instance [6, 7, 60, 61] and the references therein. Generally speaking, an observer-based fault detection system consists of an observer-based residual generator and a residual evaluator. In most of the studies, residual generator and residual evaluator are designed separately. The residual generator is so designed that, the residual signal is robust against unknown inputs and simultaneously sensitive to faults, for instance, see [31, 33–35] to list a few. For residual evaluator design, fewer attentions have been paid. Since the performance of fault detection system is decided by residual generator and evaluator together, a residual generator optimized under some performance index does not automatically result in an optimal fault detection system [6]. Therefore, in order to achieve an optimal FD performance, it is necessary to study the approach which design the residual generator and evaluator in an integrated manner instead of separate handling of these units. For the integrated design of FD system, a criterion which can evaluate the performance of the whole FD system is needed. In practice, essential requirements on an FD system are generally expressed in terms of a lower *false alarm rate* and a higher *fault detection rate*, and an optimal trade-off between them is of primary interest in the FD system design [6]. The concepts FDR and FAR are originally defined in the statistic framework and extended for FD systems with deterministic residual signals by [23, 24] in the context of a norm-based residual evaluation. An integrated design framework for linear systems is systematically introduced in [6], where concepts of $SDFA$, SDF, the simplified norm-based definition of FAR and FDR are defined. The integrated design problem has been formulated in two ways: (1) Given FDR, minimizing the $SDFA$; (2) Given FAR, maximizing the SDF [6].

On the other hand, since most practical systems are nonlinear in nature, observer-based FD techniques for nonlinear systems have also received a lot of attention, see for instance [12, 62–66]. As there is no uniform solution for general nonlinear systems, most of the time, FD system design for special kinds of nonlinear systems is studied. Among them, Lipschitz nonlinear systems is one of the most important classes of nonlinear systems, since under some conditions, more general nonlinear systems can be transformed into Lipschitz nonlinear systems as discussed in [15]. Different FD approaches for Lipschitz nonlinear systems have been proposed in [16–18, 56–59] etc. which mainly focus on residual generator design. As mentioned above, an integrated design approach is needed to achieve the optimal performance of the whole FD system. In this chapter, the integrated design approach is extended to uncertain Lipschitz nonlinear systems. In practice, usually the requirement of fault detectability should be first fulfilled, so the following integrated design formulation is used: given FDR, minimizing the $SDFA$.

3.2 Preliminaries

Consider the following uncertain nonlinear systems

$$\Sigma_{\mathscr{S}} : \begin{cases} \dot{x} &= \bar{A}x + \phi(x,u) + \bar{B}u + \bar{E}_d d + E_f f \\ y &= \bar{C}x + \bar{D}u + \bar{F}_d d + F_f f \end{cases} \tag{3.1}$$

3.2 Preliminaries

where $x \in \mathfrak{R}^n$ is the state vector, $u \in \mathfrak{R}^m$ is the input vector, $y \in \mathfrak{R}^p$ is the output vector, $f \in \mathfrak{R}^l$ is the fault vector to be detected, and $d \in \mathfrak{R}^q$ is the unknown input vector. Moreover, matrices $\bar{A}$, $\bar{B}$, $\bar{E}_d$, $\bar{C}$, $\bar{D}$, $\bar{F}_d$ in (3.1) are uncertain of the form $\bar{X} = X + \Delta X$, where $X \in \{A, B, E_d, C, D, F_d\}$ is known matrix with appropriate dimension. Similarly, matrices E_f and F_f are also known. $\Delta X \in \{\Delta A, \Delta B, \Delta E_d, \Delta C, \Delta D, \Delta F_d\}$ is norm bounded uncertainty and can be expressed as

$$\begin{bmatrix} \Delta A & \Delta B & \Delta E_d \\ \Delta C & \Delta D & \Delta F_d \end{bmatrix} = \begin{bmatrix} E \\ F \end{bmatrix} \Delta(t) \begin{bmatrix} G & H & K \end{bmatrix}$$

where E, F, G, H, K are known matrices with appropriate dimensions and $\Delta(t)$ is bounded by

$$\Delta(t)^T \Delta(t) \leq I.$$

The nonlinear function $\phi(x, u)$ is assumed to be Lipschitz in x with a Lipschitz constant $\gamma \geq 0$. i.e $\forall x, \hat{x}, u$

$$\|\phi(x, u) - \phi(\hat{x}, u)\| \leq \gamma \|x - \hat{x}\|.$$

Here $\|.\|$ denotes the Euclidean norm of a vector. In addition the following assumptions should be made throughout:

1. $A + \Delta A$ is asymptotically stable for all ΔA;

2. (C, A) is detectable.

The first assumption can be checked by standard Lyapunov approach with LMI tools [16]. The nonlinear observer based fault detection filter is designed as

$$\Sigma_{\mathscr{F}} : \begin{cases} \dot{\hat{x}} = A\hat{x} + \phi(\hat{x}, u) + Bu + L(y - C\hat{x} - Du) \\ r = y - C\hat{x} - Du \end{cases} \quad (3.2)$$

where $r \in \mathfrak{R}^p$ is the residual signal, $L \in \mathfrak{R}^{n \times p}$ is the observer gain. Denoting the estimation error $e = x - \hat{x}$, then we have the following observer error dynamics

$$\Sigma_{\mathscr{E}} : \begin{cases} \dot{e} = (A - LC)e + \Psi + (\Delta A - L\Delta C)x \\ \quad + (\Delta B - L\Delta D)u + (\bar{E}_d - L\bar{F}_d)d \\ \quad + (E_f - LF_f)f \\ r = Ce + \Delta Cx + \Delta Du + \bar{F}_d d + F_f f \end{cases} \quad (3.3)$$

with $\Psi = \phi(x,u) - \phi(\hat{x},u)$. Combining system (3.1) and the error dynamics (3.3), the residual generator dynamics is as follows

$$\Sigma_{\mathscr{R}}: \begin{cases} \dot{x}_0 = \bar{A}_0 x_0 + \Psi_0 + \bar{E}_0 d_0 + E_{0,f} f \\ r = \bar{C}_0 x_0 + \bar{F}_0 d_0 + F_{0,f} f \end{cases} \qquad (3.4)$$

with

$$x_0 = \begin{bmatrix} x \\ e \end{bmatrix}; d_0 = \begin{bmatrix} u \\ d \end{bmatrix};$$

$$\begin{bmatrix} \bar{A}_0 & \bar{E}_0 \\ \bar{C}_0 & \bar{F}_0 \end{bmatrix} = \begin{bmatrix} A_0 & E_0 \\ C_0 & F_0 \end{bmatrix} + \begin{bmatrix} \Delta A_0 & \Delta E_0 \\ \Delta C_0 & \Delta F_0 \end{bmatrix};$$

$$A_0 = \begin{bmatrix} A & 0 \\ 0 & A-LC \end{bmatrix}; E_0 = \begin{bmatrix} B & E_d \\ 0 & E_d - LF_d \end{bmatrix};$$

$$E_{0,f} = \begin{bmatrix} E_f \\ E_f - LF_f \end{bmatrix}; \Psi_0 = \begin{bmatrix} \phi(x,u) \\ \phi(x,u) - \phi(\hat{x},u) \end{bmatrix};$$

$$C_0 = \begin{bmatrix} 0 & C \end{bmatrix}; F_0 = \begin{bmatrix} 0 & F_d \end{bmatrix}; F_{0,f} = F_f;$$

$$\Delta A_0 = \begin{bmatrix} \Delta A & 0 \\ \Delta A - L\Delta C & 0 \end{bmatrix}; \Delta C_0 = \begin{bmatrix} 0 & \Delta C \end{bmatrix};$$

$$\Delta E_0 = \begin{bmatrix} \Delta B & \Delta E_d \\ \Delta B - L\Delta D & \Delta E_d - L\Delta F_d \end{bmatrix};$$

$$\Delta F_0 = \begin{bmatrix} \Delta D & \Delta F_d \end{bmatrix};$$

$$\begin{bmatrix} \Delta A_0 & \Delta E_0 \\ \Delta C_0 & \Delta F_0 \end{bmatrix} = \begin{bmatrix} \bar{E} \\ \bar{F} \end{bmatrix} \Delta(t) \begin{bmatrix} \bar{G} & \bar{H} \end{bmatrix};$$

$$\bar{E} = \begin{bmatrix} E^T & (E-LF)^T \end{bmatrix}^T; \bar{F} = F$$

$$\bar{G} = \begin{bmatrix} G & 0 \end{bmatrix}; \bar{H} = \begin{bmatrix} H & K \end{bmatrix}.$$

In control theory, the $\mathcal{L}_2$ norm is widely used to measure the "energy" of a signal, which is defined as follows.

3.2 Preliminaries

Definition 3.2.1. *The $\mathcal{L}_2$ norm of a vector-valued signal u(t) is defined by ([6])*

$$\|u\|_2 = (\int_0^\infty u^T(t)u(t)dt)^{1/2}.$$

In practice, the $\mathcal{L}_2$ norm is usually computed in a time window $[0,\tau]$ as

$$\|u\|_{2,[0,\tau]} = (\int_0^\tau u^T(t)u(t)dt)^{1/2}.$$

Using the $\mathcal{L}_2$ norm, d_0 in (3.4) is assumed to be bounded by

$$\|d_0\|_2 \leq \delta_{d,max}. \tag{3.5}$$

For the purpose of residual evaluation, the $\mathcal{L}_2$ norm of the residual signal is adopted as evaluation function:

$$J = \|r\|_2.$$

The decision logic of fault detection is as follows:

$$J > J_{th} \Longrightarrow faulty$$
$$J \leq J_{th} \Longrightarrow fault-free$$

where J_{th} is the threshold. A false alarm is created if

$$J > J_{th} \quad \text{for} \quad f = 0,$$

and a fault is detected if

$$J > J_{th} \quad \text{for} \quad f \neq 0.$$

In order to measure the influence of the faults to the residual signal r in (3.4), H_- index is defined.

Definition 3.2.2. *[67] Given the system $\Sigma_\mathcal{R}$ (3.4), assume that $d_0 = 0$, then the H_- index can be defined as*

$$\|\Sigma_\mathcal{R}\|_- = \inf_{f \neq 0} \frac{\|r\|_2}{\|f\|_2}.$$

For H_- index to be larger than some positive number β can be defined as

$$\|\Sigma_{\mathscr{R}}\|_- = \inf_{f\neq 0} \frac{\|r\|_2}{\|f\|_2} \geq \beta,$$

or

$$\|r\|_2 \geq \beta \|f\|_2. \tag{3.6}$$

$\beta > 0$ which fulfills (3.6) is called H_- gain of system $\Sigma_{\mathscr{R}}$.

The following lemma is very useful in the sequel study.

Lemma 3.2.1. *[68] Let G, Q, E, $F(t)$ be real matrices of appropriate dimensions with $F(t)$ being a matix function and $F(t)^T F(t) \leq I$, then for any scalar $\epsilon > 0$,*

$$MF(t)E + E^T F(t) M^T \leq \frac{1}{\epsilon} MM^T + \epsilon E^T E.$$

3.3 Definitions of FAR and FDR

3.3.1 Review of the definitions in the statistical framework

FAR and FDR are two concepts that are originally defined in the statistical framework [42, 44]. Considering a stochastic process corrupted with the unknown input vector d and the fault vector f, the definitions of FAR and FDR are as follows:

Definition 3.3.1. *The probability FAR*

$$FAR = prob(J > J_{th}|f=0)$$

is called false alarm rate.

Definition 3.3.2. *The probability FDR*

$$FDR = prob(J > J_{th}|f \neq 0)$$

is called fault detection rate.

3.3.2 Definitions in the norm-based framework

For deterministic system (3.1), new definitions are needed. In the context of a norm based residual evaluation, the following definitions are introduced in [6].

Definition 3.3.3. *Given residual generator $\Sigma_\mathscr{R}$ and threshold J_{th}, the set SDF defined by*

$$SDF = \{f | J > J_{th} \text{ for } f \neq 0\}$$

is called set of detectable faults.

The size of SDF is a direct measurement of the FD system performance regarding to the fault detectability. Since it is very difficult to express FDR in terms of the size of SDF, a simplified definition of FDR is introduced.

Definition 3.3.4. *FDR given by*

$$FDR = \frac{\beta \delta_{f,min}}{J_{th}}$$

is called FDR in the norm based framework. Where $\beta > 0$ is the H_- gain from faults to residual signal in the case that there are no disturbances and uncertainties ($d_0 = 0$, $\Delta = 0$), and $\delta_{f,min} > 0$ is the minimum size of the f vector which is defined as faults to be detected.

The physical meaning of this definition is that, the larger the faults are ($\delta_{f,min}$ is larger), the larger FDR is; the larger the threshold is (J_{th} is larger), the smaller FDR will be.

Definition 3.3.5. *Given residual generator $\Sigma_\mathscr{R}$ and threshold J_{th}, the set $SDFA$ defined by*

$$SDFA = \{d | J > J_{th} \text{ for } f = 0\}$$

is called set of disturbances that cause false alarms.

The size of $SDFA$ indicates the number of the possible false alarms, and thus builds a direct measurement of the FD system performance regarding to the intensity of false alarms.

3.4 Integrated design of FD systems

3.4.1 Problem formulation

The objective of the integrated design of FD system is to achieve an optimal trade-off between fault detectability and false alarm number. Since in practice, usually the requirement of fault detectability should be first fulfilled, so the integrated design problem is formulated as: given FDR, minimizing the false alarm number ($SDFA$).

The design parameters of the FD system are the observer gain L in (3.4) and the threshold J_{th}. As FDR is given, according to Definition 3.3.4, threshold J_{th} should be set as

$$J_{th} = \beta \xi_{fd} \tag{3.7}$$

with

$$\xi_{fd} = \frac{\delta_{f,min}}{FDR}, \quad 0 < FDR \leq 1$$

and β is the H_- gain from faults to residual signal in the case that there are no disturbances and uncertainties. With $d_0 = 0$ and $\Delta = 0$, residual generator dynamics (3.4) becomes

$$\begin{cases} \dot{x}_f &= (A - LC)x_f + \Psi + (E_f - LF_f)f \\ r_f &= Cx_f + F_f f. \end{cases} \tag{3.8}$$

β fulfills

$$\|r_f\|_2 \geq \beta \|f\|_2.$$

$\delta_{f,min}$ is the minimum size of the f vector which is defined as faults to be detected, we have

$$\|f\|_2 \geq \delta_{f,min}.$$

By setting J_{th} as in (3.7), the requirement of FDR is fulfilled. In the following we will try to minimize the false alarm number. Since false alarms are created in the fault-free case with $f = 0$, in this case the residual generator dynamics (3.4) becomes

$$\begin{cases} \dot{x}_{0,d} &= \bar{A}_0 x_{0,d} + \Psi_0 + \bar{E}_0 d_0 \\ r_d &= \bar{C}_0 x_{0,d} + \bar{F}_0 d_0. \end{cases} \tag{3.9}$$

According to Definition 3.3.5, for residual generator (3.9) and threshold (3.7), $SDFA$ is as

$$SDFA = \{d_0| \ \|r_d\|_2 > \beta \xi_{fd}\}.$$

In the norm-based framework, FAR ($0 \leq FAR \leq 1$) can be defined as the size of $SDFA$. False alarms are created when evaluated residual driven by disturbances becomes larger than the threshold in the fault-free case. In the norm-based framework, the size of disturbances is measured by its $\mathcal{L}_2$ norm. For residual generator dynamics (3.9), the considered disturbances are bounded by $\|d_0\|_2 \leq \delta_{d,max}$. Since the upper bound of disturbances is fixed, FAR can be

3.4 Integrated design of FD systems

defined based on the lower bound of disturbances which leads to false alarms. This lower bound is denoted as $\delta_{fa,min}$, we have

$$\|d_0\|_2 > \delta_{fa,min} \tag{3.10}$$

$$\iff \|r_d\|_2 - \beta \xi_{fd} > 0. \tag{3.11}$$

In order to achieve a smaller FAR, $\delta_{fa,min}$ should be larger. This relation can be represented by (considering the range $0 \leq FAR \leq 1$)

$$\delta_{fa,min} = (1 - FAR)\delta_{d,max}. \tag{3.12}$$

Note that, when $FAR = 0$, (3.10) will never be fulfilled since d_0 is bounded by $\delta_{d,max}$, which leads to the smallest size of $SDFA$; and when $FAR = 1$, (3.10) is almost always true, which leads to the largest size of $SDFA$.

Based on (3.10), (3.11) and (3.12), the integrated design problem is formulated as: given FDR, finding observer gain L so that FAR is minimized.

3.4.2 A solution to the integrated design problem

With a given FDR, (3.11) can be transformed into

$$\frac{\|r_d\|_2}{\beta \xi_{fd}} > 1,$$

then based on it, one sufficient condition for (3.10) is

$$\|d\|_2 \geq (1 - FAR)\delta_{d,max} \frac{\|r_d\|_2}{\beta \xi_{fd}}. \tag{3.13}$$

In (3.13), FAR and FDR are connected, which is the key step for the optimization. Because (3.13) is a sufficient condition, for the given FDR, the FAR calculated by (3.13) is larger or equal to its actual value. So a suboptimal solution can be achieved by minimizing FAR based on (3.13). The range of FAR has been defined as $0 \leq FAR \leq 1$, in the following study, it is assumed that

$$0 \leq FAR < 1.$$

Since FAR should be minimized, this assumption will not lead to a conservative result. (3.13) can be transformed into

$$\|r_d\|_2 - \beta \xi_{fd} \frac{\|d\|_2}{(1 - FAR)\delta_{d,max}} \leq 0 \tag{3.14}$$

Chapter 3 Observer-based FD for Lipschitz nonlinear systems: an integrated desgin approach

where β fulfills

$$\|r_f\|_2 \geq \beta \|f\|_2. \tag{3.15}$$

Based on (3.14) and (3.15), for a given FDR, the minimization of FAR can be achieved in an iterative way as follows:

Step I. Set the initial value of FAR.

Step II. If there exist β and observer gain L, which fulfill (3.14) and (3.15), then decrease (otherwise increase) the value of FAR until we get the minimum FAR.

The above algorithm gives an elegant tool for the minimization of FAR provided FDR is given. Now the question is how to check the existence of β and L in Step II. To this end, the following theorem is proposed.

Theorem 3.4.1. *Given residual generator dynamics (3.8) and (3.9), assume that $x_f(0) = 0$, $x_{0,d}(0) = 0$. Then*

$$\|r_d\|_2 - \beta\lambda\|d\|_2 \leq 0 \tag{3.16}$$

$$\|r_f\|_2 \geq \beta\|f\|_2 \tag{3.17}$$

with

$$\lambda = \frac{\xi_{fd}}{(1 - FAR)\delta_{d,max}},$$

if there exist some $\epsilon > 0$, $\beta > 0$, L and symmetric matrices $P_1 > 0$, $P_2 > 0$, $Q \leq 0$ so that

$$\begin{bmatrix} \Omega_1 & \Omega_2 \\ * & \Omega_3 \end{bmatrix} \leq 0 \tag{3.18}$$

$$\begin{bmatrix} N_5 & Q(E_f - LF_f) + C^T F_f & \gamma Q \\ * & F_f^T F_f - \beta^2 I & 0 \\ * & * & I \end{bmatrix} \geq 0 \tag{3.19}$$

where

$$\Omega_1 = \begin{bmatrix} N_1 & 0 & N_3 & P_1 E_d + \varepsilon G^T K \\ * & N_2 & 0 & P_2(E_d - LF_d) \\ * & * & N_4 & \varepsilon H^T K \\ * & * & * & -\beta^2 \lambda^2 I + \varepsilon K^T K \end{bmatrix}$$

$$\Omega_2 = \begin{bmatrix} 0 & P_1 E & 0 & \gamma P_1 \\ C^T & P_2(E - LF) & \gamma P_2 & 0 \\ 0 & 0 & 0 & 0 \\ F_d^T & 0 & 0 & 0 \end{bmatrix}$$

$$\Omega_3 = \begin{bmatrix} -I & -F & 0 & 0 \\ * & -\varepsilon I & 0 & 0 \\ * & * & -I & 0 \\ * & * & * & -I \end{bmatrix}$$

$$\begin{aligned} N_1 &= P_1 A + A^T P_1 + \varepsilon G^T G + I \\ N_2 &= P_2(A - LC) + (A - LC)^T P_2 + I \\ N_3 &= P_1 B + \varepsilon G^T H \\ N_4 &= -\beta^2 \lambda^2 I + \varepsilon H^T H \\ N_5 &= Q(A - LC) + (A - LC)^T Q - I + C^T C \end{aligned}$$

Proof. The proof includes two parts.

Part 1: Let
$$V_d = x_{0,d}^T P x_{0,d}, \quad P = \begin{bmatrix} P_1 & 0 \\ 0 & P_2 \end{bmatrix} > 0.$$

It holds

$$r_d^T r_d - \beta^2 \lambda^2 d^T d + \dot{V}_d \leq 0 \qquad (3.20)$$
$$\implies \int_0^\infty r_d^T r_d - \beta^2 \lambda^2 \int_0^\infty d^T d + V_d(\infty) \leq 0$$
$$\implies \|r_d\|_2 - \beta \lambda \|d\|_2 \leq 0.$$

Hence (3.20) is the sufficient condition for (3.16). We have

$$\dot{V}_d = 2x_{0,d}^T P[\bar{A}_0 x_{0,d} + \bar{E}_0 d_0] + 2x_{0,d}^T P \Psi_0. \qquad (3.21)$$

Chapter 3 Observer-based FD for Lipschitz nonlinear systems: an integrated desgin approach

---

Using Cauchy-Schwarz inequality and the Lipschitz property of Ψ_0, it turns out

$$\begin{aligned} 2x_{0,d}^T P \Psi_0 &\leq 2\|Px_{0,d}\|\|\Psi_0\| \\ &\leq 2\gamma \|Px_{0,d}\|\|x_{0,d}\| \\ &\leq \gamma^2 x_{0,d}^T PP x_{0,d} + x_{0,d}^T x_{0,d}. \end{aligned} \qquad (3.22)$$

Substituting (3.22) into (3.21) yields

$$\begin{aligned} \dot{V}_d &\leq 2x_{0,d}^T P[\bar{A}_0 x_{0,d} + \bar{E}_0 d_0] \\ &\quad + \gamma^2 x_{0,d}^T PP x_{0,d} + x_{0,d}^T x_{0,d}. \end{aligned}$$

Based on it, a sufficient condition for (3.20) is

$$(\bar{C}_0 x_{0,d} + \bar{F}_0 d_0)^T (\bar{C}_0 x_{0,d} + \bar{F}_0 d_0) - \beta^2 \lambda^2 d^T d$$
$$+ 2x_{0,d}^T P[\bar{A}_0 x_{0,d} + \bar{E}_0 d_0] + \gamma^2 x_{0,d}^T PP x_{0,d}$$
$$+ x_{0,d}^T x_{0,d} \leq 0$$
$$\iff \begin{bmatrix} x_{0,d} \\ d_0 \end{bmatrix}^T \chi_1 \begin{bmatrix} x_{0,d} \\ d_0 \end{bmatrix} \leq 0 \qquad (3.23)$$

where

$$\chi_1 = \begin{bmatrix} \bar{C}_0^T \\ \bar{F}_0^T \end{bmatrix} \begin{bmatrix} \bar{C}_0 & \bar{F}_0 \end{bmatrix} + \begin{bmatrix} N_6 & P\bar{E}_0 \\ * & -\beta^2\lambda^2 I \end{bmatrix},$$

and

$$N_6 = P\bar{A}_0 + \bar{A}_0^T P + \gamma^2 PP + I.$$

Therefore

$$\chi_1 \leq 0 \qquad (3.24)$$
$$\implies \|r_d\|_2 - \beta\lambda\|d\|_2 \leq 0.$$

3.4 Integrated design of FD systems

Applying the Schur complement we can rewrite (3.24) into

$$\begin{bmatrix} N_6 & P\bar{E}_0 & \bar{C}_0^T \\ \bar{E}_0^T P & -\beta^2\lambda^2 I & \bar{F}_0^T \\ \bar{C}_0 & \bar{F}_0 & -I \end{bmatrix} \leq 0 \iff \qquad (3.25)$$

$$\begin{bmatrix} PA_0 + A_0^T P + \gamma^2 PP + I & P E_0 & C_0^T \\ E_0^T P & -\beta^2\lambda^2 I & F_0^T \\ C_0 & F_0 & -I \end{bmatrix}$$

$$+ \begin{bmatrix} P\Delta A_0 + \Delta A_0^T P & P\Delta E_0 & \Delta C_0^T \\ \Delta E_0^T P & 0 & \Delta F_0^T \\ \Delta C_0 & \Delta F_0 & 0 \end{bmatrix} \leq 0.$$

Split the second matrix in the above inequality into

$$\begin{bmatrix} P\Delta A_0 + \Delta A_0^T P & P\Delta E_0 & \Delta C_0^T \\ \Delta E_0^T P & 0 & \Delta F_0^T \\ \Delta C_0 & \Delta F_0 & 0 \end{bmatrix} = \chi_2 + \chi_2^T$$

where

$$\chi_2 = \begin{bmatrix} P\bar{E} \\ 0 \\ F \end{bmatrix} \Delta(t) \begin{bmatrix} \bar{G} & \bar{H} & 0 \end{bmatrix}.$$

Then according to Lemma 3.2.1, (3.25) holds if there exists an $\varepsilon > 0$ such that

$$\begin{bmatrix} N_7 & PE_0 & C_0^T \\ E_0^T P & -\beta^2\lambda^2 I & F_0^T \\ C_0 & F_0 & -I \end{bmatrix} + \frac{1}{\varepsilon} \begin{bmatrix} P\bar{E} \\ 0 \\ F \end{bmatrix} \begin{bmatrix} P\bar{E} \\ 0 \\ F \end{bmatrix}^T$$

$$+ \varepsilon \begin{bmatrix} \bar{G} & \bar{H} & 0 \end{bmatrix}^T \begin{bmatrix} \bar{G} & \bar{H} & 0 \end{bmatrix} \leq 0$$

with $N_7 = PA_0 + A_0^T P + \gamma^2 PP + I$. Applying the Schur complement yields

$$\begin{bmatrix} N_8 & PE_0 + \varepsilon \bar{G}^T \bar{H} & C_0^T & P\bar{E} \\ * & -\beta^2\lambda^2 I + \varepsilon \bar{H}^T \bar{H} & F_0^T & 0 \\ * & * & -I & F \\ * & * & * & -\varepsilon I \end{bmatrix} \leq 0$$

with $N_8 = PA_0 + A_0^T P + \gamma^2 PP + I + \varepsilon \bar{G}^T \bar{G}$. Substituting $P = \begin{bmatrix} P_1 & 0 \\ 0 & P_2 \end{bmatrix}$ into the above inequality, it turns out

$$\begin{bmatrix} N_9 & 0 & N_{10} & N_{11} & 0 & P_1 E \\ * & N_{12} & 0 & N_{13} & C^T & P_2(E-LF) \\ * & * & N_{14} & \varepsilon H^T K & 0 & 0 \\ * & * & * & N_{15} & F_d^T & 0 \\ * & * & * & * & -I & -F \\ * & * & * & * & * & -\varepsilon I \end{bmatrix} \leq 0$$

where

$$\begin{aligned} N_9 &= P_1 A + A^T P_1 + \varepsilon G^T G + \gamma^2 P_1 P_1 + I \\ N_{10} &= P_1 B + \varepsilon G^T H, N_{11} = P_1 E_d + \varepsilon G^T K \\ N_{12} &= P_2(A-LC) + (A-LC)^T P_2 + \gamma^2 P_2 P_2 + I \\ N_{13} &= P_2(E_d - LF_d), N_{14} = -\beta^2 \lambda^2 I + \varepsilon H^T H \\ N_{15} &= -\beta^2 \lambda^2 I + \varepsilon K^T K. \end{aligned}$$

Finally, applying the Schur complement again, we have (3.18) of Theorem 3.4.1, which is the sufficient condition for (3.16).

Part 2: Let

$$V_f(x) = x_f^T Q x_f, \quad Q \leq 0.$$

It holds

$$r_f^T r_f - \beta^2 f^T f + \dot{V}_f \geq 0 \tag{3.26}$$

$$\implies \int_0^\infty r_f^T r_f - \beta^2 \int_0^\infty f^T f + V_f(\infty) \geq 0$$

$$\implies \|r_f\|_2 \geq \beta \|f\|_2.$$

So (3.26) is the sufficient condition for (3.17). We have

$$\begin{aligned} \dot{V}_f &= 2x_f^T Q[(A-LC)x_f + (E_f - LF_f)f] \\ &\quad + 2x_f^T Q \Psi. \end{aligned} \tag{3.27}$$

3.4 Integrated design of FD systems

Using Cauchy-Schwarz inequality and the Lipschitz property of Ψ, it turns out

$$\begin{aligned} 2x_f^T Q\Psi &\geq -2\|Qx_f\|\|\Psi\| \\ &\geq -2\gamma\|Qx_f\|\|x_f\| \\ &\geq -\gamma^2 x_f^T QQ x_f - x_f^T x_f. \end{aligned} \quad (3.28)$$

Substituting (3.28) into (3.27) yields

$$\begin{aligned} \dot{V}_f &\geq 2x_f^T Q[(A-LC)x_f + (E_f - LF_f)f] \\ &\quad -\gamma^2 x_f^T QQ x_f - x_f^T x_f. \end{aligned}$$

Based on it, a sufficient condition for (3.26) is

$$(Cx_f + F_f f)^T (Cx_f + F_f f) - \beta^2 f^T f$$
$$+ 2x_f^T Q(A-LC)x_f + 2x_f^T Q(E_f - LF_f)f$$
$$-\gamma^2 x_f^T QQ x_f - x_f^T x_f \geq 0$$
$$\iff \begin{bmatrix} x_f \\ f \end{bmatrix}^T \begin{bmatrix} N_{16} & N_{17} \\ * & F_f^T F_f - \beta^2 I \end{bmatrix} \begin{bmatrix} x_f \\ f \end{bmatrix} \geq 0$$

with

$$\begin{aligned} N_{16} &= Q(A-LC) + (A-LC)^T Q - \gamma^2 QQ \\ &\quad - I + C^T C \\ N_{17} &= Q(E_f - LF_f) + C^T F_f. \end{aligned}$$

Applying the Schur complement yields (3.19) of Theorem 3.4.1, which is the sufficient condition for (3.17). This completes the proof. □

In Theorem 3.4.1, (3.18) and (3.19) are nonlinear matrix inequalities (NMIs) which can be transformed into standard LMIs by setting

$$Q = -P_2, \ Y = P_2 L.$$

So the integrated design problem can be solved by powerful LMI tools.

29

3.5 A design example

3.5.1 System description

Consider the FD problem of a system in the form of (3.1) with coefficient matrices as

$$A = \begin{bmatrix} -6.5 & 3.9 & 5.2 \\ 0 & -9.1 & 3.9 \\ 1.3 & 3.9 & -7.8 \end{bmatrix}, \quad B = \begin{bmatrix} 1 \\ 2 \\ 1.5 \end{bmatrix},$$

$$C = \begin{bmatrix} 1 & 2 & -1 \\ 2 & -1 & 3 \end{bmatrix}, \quad D = \begin{bmatrix} 0.5 \\ 0.3 \end{bmatrix},$$

$$E_d = \begin{bmatrix} -0.3 & 1 & 0.6 \\ 0 & 0.3 & 0.5 \\ 0.4 & 0 & -0.2 \end{bmatrix},$$

$$E_f = \begin{bmatrix} 1.3 & 0.65 \\ -0.39 & 1.04 \\ 0.78 & -1.17 \end{bmatrix}, \quad F = \begin{bmatrix} 0.35 \\ 0.1 \end{bmatrix},$$

$$F_d = \begin{bmatrix} 0.7 & 1 & -0.3 \\ 0 & 0.6 & 0.2 \end{bmatrix}, \quad F_f = \begin{bmatrix} 1.6 & 0 \\ 0 & -1.6 \end{bmatrix},$$

$$E = \begin{bmatrix} 0.2 \\ 0.3 \\ 0.15 \end{bmatrix}, \quad \phi(x,u) = 0.5 \begin{bmatrix} \sin(x_1) \\ \cos(x_2) \\ 0 \end{bmatrix},$$

$$G = \begin{bmatrix} 0.25 & 0.1 & 0.33 \end{bmatrix}, \quad H = 0.12,$$

$$K = \begin{bmatrix} 0.16 & 0.23 & 0.31 \end{bmatrix}.$$

In the integrated design, set

$$FDR = 1,$$

then following the algorithm in Section 3.4.2 to minimize the FAR, we get

$$FAR_{min} = 0.12,$$

and the optimal observer gain matrix is

$$L_{opt} = \begin{bmatrix} 0.82 & -0.32 \\ -0.14 & -0.41 \\ 0.34 & 0.52 \end{bmatrix}.$$

The H_- gain from faults to residual signal of (3.8) is $\beta = 1.5$. The input and the disturbances are set as

$$u(t) = 3, \quad d(t) = \begin{bmatrix} 2.8cos(10t) & 2 & 4sin(10t) \end{bmatrix}. \tag{3.29}$$

The simulations are carried out for $\tau = 3000$ seconds. $d_0 = \begin{bmatrix} d & u \end{bmatrix}^T$ is bounded by

$$\|d_0\|_{2,[0,\tau]} \leq \delta_{d,max} = 332.4$$

The faults which are defined to be detected is bounded as

$$\|f\|_{2,[0,\tau]} \geq \delta_{f,min} = 136.7$$

then according to (3.7), the threshold is computed as

$$J_{th} = \beta \frac{\delta_{f,min}}{FDR} = 205.6$$

3.5.2 Simulation results

Two faulty cases are considered as shown in Fig 3.1. For the first case in Fig 3.1a, Fault 1 appears at $t = 1500$ seconds as a step function $f_1 = 3$ and Fault 2 appears at $t = 1700$ seconds as a step function $f_2 = -2$. For the second case, Fault 1 and Fault 2 are ramp signals as in Fig 3.1b. In both cases, the size of the faults is

$$\|f\|_{2,[0,\tau]} = \delta_{f,min} = 136.7$$

where $f = \begin{bmatrix} f_1 & f_2 \end{bmatrix}^T$. So the size of the faults equal to the minimum value to be detected. Two faulty cases are denoted as F_1 and F_2. In the simulation results, the solid line represents the behavior of residual signals $r_i(i = 1, 2)$ or evaluated residual J, and the dashed line represents the threshold J_{th}.

Performance of FD system: Fig 3.2 shows the simulation results for the first faulty case. From Fig 3.2a we can see that, after f_1 becomes nonzero at $t = 1500$ seconds, there is a significant change in r_1, and after f_2 becomes nonzero at $t = 1700$ seconds, there is a significant change in r_2. In Fig 3.2b, after the fault appears, the evaluated residual J increases much faster and becomes larger than the threshold at $t = 1860$ seconds, which leads to a successful fault detection.

Study of FDR: According to Definition 3.3.4 of FDR, considering the case when the system is only driven by the faults i.e. $d_0 = 0$, $\Delta = 0$. Since fault detection rate is set as $FDR = 1$,

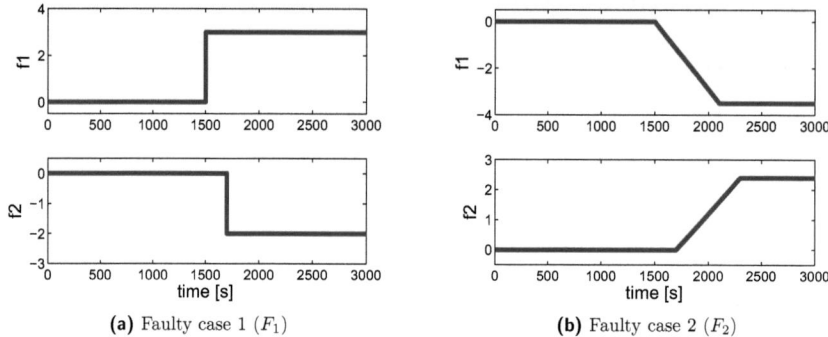

Figure 3.1: Two faulty cases

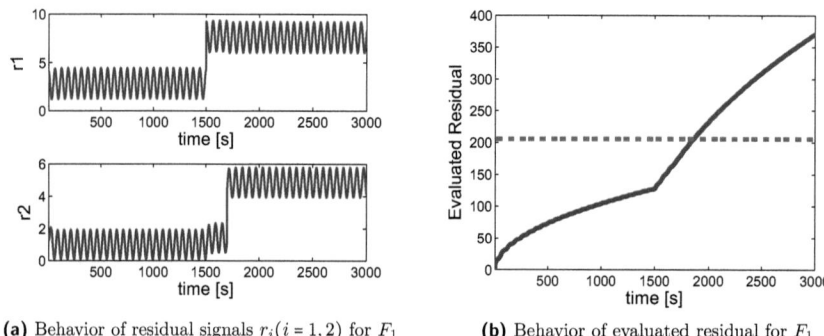

(a) Behavior of residual signals $r_i(i = 1, 2)$ for F_1 (b) Behavior of evaluated residual for F_1

Figure 3.2: Performance of FD system for F_1

it means all faults whose size larger or equal to $\delta_{f,min}$ should be detected. Fig 3.3 shows the behavior of evaluated residual J for both faulty cases. We can see that, before the faults appear, the evaluated residual J is equal to zero since $d_0 = 0$, and after the occurrence of the faults, J turns to be non-zero and finally crosses the threshold, which leads to a successful fault detection. So for both cases the requirement of FDR is fulfilled.

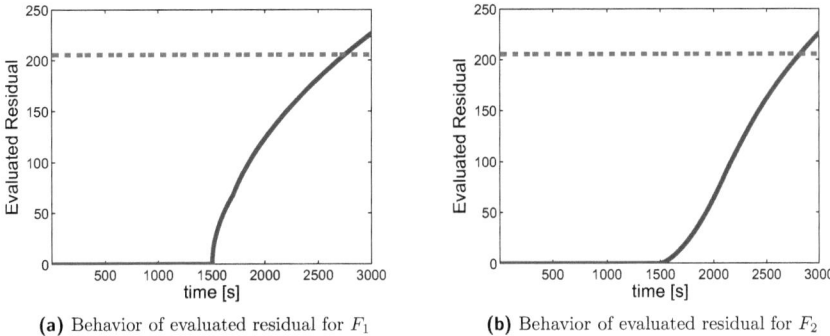

(a) Behavior of evaluated residual for F_1

(b) Behavior of evaluated residual for F_2

Figure 3.3: Behavior of evaluated residual for F_1 and F_2 when $d_0 = 0$, $\Delta = 0$

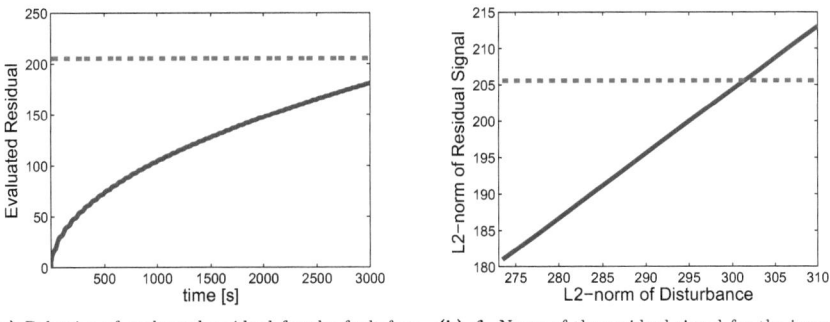

(a) Behavior of evaluated residual for the fault-free case

(b) $\mathcal{L}_2$-Norm of the residual signal for the increased $\|d_0\|_2$

Figure 3.4: Study of FAR

Study of FAR: According to the definition of FAR in (3.10) and (3.11), since we have

$FAR_{min} = 0.12$, when d_0 is bounded by

$$\|d_0\|_2 \le (1 - FAR)\delta_{d,max} = 292.5, \qquad (3.30)$$

then there should be no false alarms. Considering the fault-free case with $d_0 = \begin{bmatrix} d & u \end{bmatrix}^T$, where d and u are as in (3.29), we have

$$\|d_0\|_2 = 273.4 < 292.5 \qquad (3.31)$$

Fig 3.4a shows the simulation results. We can see that, the evaluated residual stays under the threshold, there is no false alarm. In order to validate the FAR, a series of simulations have been carried out, in which the size of disturbances is step by step increased as $\tilde{d}(t) = p*d(t)(p \ge 1)$, where $d(t)$ is as in (3.29). The $\mathcal{L}_2$-Norm of the residual signal for different size of d_0 is shown in Fig 3.4b. We can see that, the $\mathcal{L}_2$-Norm of the evaluated residual signal increases together with $\|d_0\|_2$, and it reaches the threshold when $\|d_0\|_2 = 302.2$, which leads to false alarms. Compare it with (3.30) we know that, the obtained FAR is larger than its actual value, which is caused by the sufficient conditions in the optimization.

3.6 Summary

This chapter has proposed an integrated design approach of observer-based FD for Lipschitz nonlinear systems, which results in a suboptimal trade-off between norm-based FAR and FDR. An iterative algorithm has been presented, which minimizes the FAR for a given FDR. The effectiveness of the proposed approach has been demonstrated by a design example.

Chapter 4

Energy-balance based FDI framework for passive nonlinear systems

> The objective of this chapter is to establish an energy-balance based FDI framework for passive nonlinear systems. Based on passivity, an energy balance of passive nonlinear systems is constructed, which includes the stored, dissipated and supplied energies. In the new FDI framework, faults are defined according to their different influences on system energies. The fault detection is carried out by checking the validity of the energy balance. For fault isolation, a two-step approach is proposed including: (1) Isolation of different classes of faults (energy changes); (2) Isolation of different components. For passive nonlinear systems with unknown inputs, thresholds for fault detection and isolation are designed based on the bound of "unknown-input energy". Finally, the proposed FDI schemes are illustrated by an RLC circuit example.

4.1 Introduction

In control engineering, the concept "energy" is widely used in the stability analysis and control system design. In Lyapunov theory, a function of the system states $V(x)$ is called Lyapunov function when it fulfills $V(x) \geq 0$ with equality if and only if $x = 0$. Lyapunov function $V(x)$ can be considered as a measure of the true physical energy of the system. The asymptotical stability is guaranteed when $\dot{V}(x) \leq 0$ ($\dot{V}(x) = 0$ if and only if $x = 0$), which represents the fact that, if system loses energy over time and the energy is never restored, then eventually the system must grind to a stop and reach some final resting state. In this sense, $V(x)$ is regarded as the system "energy" which can also be applied to abstract mathematical systems, economic systems or biological systems, where the original concept of energy is not applicable. For the control

Chapter 4 Energy-balance based FDI framework for passive nonlinear systems

system desgin, dynamical systems can be viewed as "energy-transformation" devices, the control problem can then be recast as finding a dynamical system and an interconnection pattern such that the overall energy function takes the desired form [69]. This "energy-shaping" approach is the essential of passivity-based control (PBC) [70–74]. Passivity is a very important concept in control theory which relates closely to system energies. The physical meaning of passivity is that, passive systems can not store more energy than that supplied by the environment outside [21]. Energy-based approaches are particularly useful in studying complex nonlinear systems by decomposing them into simpler subsystems that, upon interconnection, add up their energies to determine the full system's behavior [69]. In this chapter, an energy-balance based FDI framework for passive nonlinear systems is proposed. Compared with the classical FDI schemes for nonlinear systems, the proposed approach can be applied to a large class of nonlinear systems i.e. all kinds of passive nonlinear systems including complex systems like switched systems, hybrid systems etc. Since for many complex nonlinear systems, the mathmatical formulation of the system energies is quite simple, the on-line computation of the proposed approaches is very low which is a very important property for the industry. Moreover, it is generally much easier to establish an "energy model" than a classical input-output model which is even impossible for many complex industry processes.

In [19–21], energy equality (energy balance) of steady-state processes and energy inequality (passive energy relation) of dynamic systems have been applied for fault detection system design, where classical FDI framework is still used i.e. fault definition, fault isolation schemes etc. In this chapter, dynamic passive nonlinear systems are considered. In order to make full use of the energy properties of passive systems, a new FDI framework is proposed. Firstly, an energy balance of passive nonlinear systems is established based on passivity. Compared with the energy inequality used in [21], besides the stored and supplied energies, dissipated energy is also taken into account. Based on it, a higher fault detectability can be achieved. On the other hand, the consideration of dissipated energy also makes it possible to isolate the faults in energy-dissipating components. Secondly, different from the classical FDI framework, the faults are defined according to their different influences on system energies. Since the change of the system energies can be detected by the energy balance, it is nature to define the faults based on different energy changes which has a very clear physical meaning. Thirdly, energy-balance based fault isolation schemes are proposed, which include two steps. The first step is to find out which kind of energy change (fault) appears in the system. After the type of the fault is isolated, the second step is to isolate the location of the fault which is to find out the faulty component. In practice, unknown inputs (disturbances) are inevitable. The influence of unknown inputs to the energy balance can be modeled as "unknown-input energy". For the fault detection and isolation of passive system with unknown inputs, thresholds are designed based on the bound of unknown-input energy.

4.2 Preliminaries

Definition 4.2.1. *[75] If there exists a function $V(x) \geq 0$, called the storage function, such that for all $x(0)$, all $\tau \geq 0$, and all functions u*

$$V(x(\tau)) - V(x(0)) \leq \int_0^\tau y^T u \, dt, \tag{4.1}$$

then the system with input u and output y is passive.

The definition of passive systems has strong physical meanings. $V(x(\tau))$ represents the energy stored in the system at time τ and $\int_0^\tau y^T u \, dt$ represents the supplied energy from the outside.

Consider an RLC circuit example as in Fig 4.1, which consists of a resistor, an inductor, a capacitor and a power source. The RLC circuit can be modeled as a passive system whose input is the voltage of the source u and output is the current $y = i$. In this case, $\int_0^\tau y^T u \, dt$ in (4.1) is exactly the power flowing into the circuit, and $V(x(\tau))$ is the energy stored in capacitor and inductor [22]. Property (4.1) expresses the fact that the "stored energy" $V(x(\tau))$ of system at any future time τ is at most equal to the sum of the stored energy $V(x(0))$ at present time and the total externally supplied energy $\int_0^\tau y^T u \, dt$ during the time interval $[0 \ \tau]$. Hence, there can be no internal "creation of energy", only internal dissipation of energy is possible [75].

4.3 Energy balance of passive systems

Based on passive relation (4.1), an energy balance of passive systems can be constructed. The internal energy dissipation $\mathscr{E}_{dis}$ is defined as

$$\mathscr{E}_{dis} = \int_0^\tau y^T u \, dt - (V(x(\tau)) - V(x(0))). \tag{4.2}$$

According to (4.1) we have

$$\mathscr{E}_{dis} \geq 0.$$

Chapter 4 Energy-balance based FDI framework for passive nonlinear systems

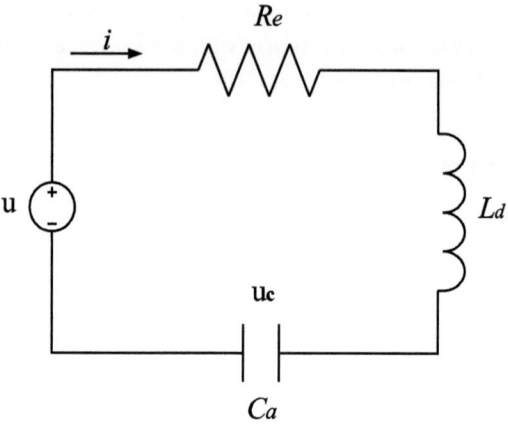

Figure 4.1: RLC circuit

Since $\mathscr{E}_{dis}$ is the sum of the dissipated energy in the time interval $[0, \tau]$, it has the following form

$$\mathscr{E}_{dis} = \int_0^\tau d(x)dt, \quad d(x) \geq 0. \tag{4.3}$$

For the RLC circuit example we have

$$d(x) = i^2 R_e$$

where i is the current and R_e is the resistance.

The energy balance of passive systems is represented by (4.2) which can be transformed into

$$V(x(\tau)) - V(x(0)) + \mathscr{E}_{dis} = \int_0^\tau y^T u \, dt. \tag{4.4}$$

Assuming the zero initial condition (i.e. $V(x(0)) = 0$), the energy balance (4.4) can be written as

$$\mathscr{H} = 0 \tag{4.5}$$

with

$$\mathcal{H} = \mathcal{E}_{stor} + \mathcal{E}_{dis} - \mathcal{E}_{sup} \tag{4.6}$$

$$\mathcal{E}_{stor} = V(x(\tau)) \tag{4.7}$$

$$\mathcal{E}_{dis} = \int_0^\tau d(x)dt \tag{4.8}$$

$$\mathcal{E}_{sup} = \int_0^\tau y^T u\, dt. \tag{4.9}$$

$\mathcal{E}_{stor}$ and $\mathcal{E}_{sup}$ are stored and supplied energies.

4.4 Definition of faults

When system faults occur, the stored energy in the system and dissipated energy out of the system can be changed. In the energy-balance based FDI framework, faults are defined according to their influences on these two kinds of energies.

4.4.1 Stored-energy change

This kind of fault is caused by the change of capacity of energy-storing components, for example the change of the capacitance of the capacitor in an RLC circuit. In this case, the stored energy becomes

$$\tilde{\mathcal{E}}_{stor} = \tilde{V}(x(\tau))$$

where $\tilde{V}(x(\tau))$ represents the changed storage function in the faulty case. When only the faults in energy-storing components are considered, the dissipated and supplied energies are the same as in (4.8) and (4.9). In this case, the new system energy balance is as

$$\tilde{\mathcal{E}}_{stor} + \mathcal{E}_{dis} = \mathcal{E}_{sup}. \tag{4.10}$$

(4.10) is equal to

$$\mathcal{E}_{dis} - \mathcal{E}_{sup} = -\tilde{\mathcal{E}}_{stor}. \tag{4.11}$$

Substituting (4.11) into (4.6) results in

$$\mathcal{H} = \Delta \mathcal{E}_{stor} \tag{4.12}$$

with $\mathcal{H}$ as in (4.6) and

$$\Delta\mathcal{E}_{stor} = \mathcal{E}_{stor} - \tilde{\mathcal{E}}_{stor} = \tilde{V}(x(\tau)) - V(x(\tau)) \qquad (4.13)$$

which is the stored-energy change.

4.4.2 Dissipated-energy change

This kind of fault is caused by the change of the dissipation rate of energy-dissipating components, for example the change of the resistance of resistor in an RLC circuit. In this case dissipated energy becomes

$$\tilde{\mathcal{E}}_{dis} = \int_0^\tau \tilde{d}(x)dt \qquad (4.14)$$

where $\tilde{d}(x)$ represents the changed dissipating function in faulty case. The faults which lead to a larger (smaller) dissipation rate are defined as

$$\begin{cases} d(x) - \tilde{d}(x) \leq 0 \text{ for a larger dissipation rate} \\ d(x) - \tilde{d}(x) \geq 0 \text{ for a smaller dissipation rate.} \end{cases} \qquad (4.15)$$

When only faults in energy-dissipating components are considered, the stored and supplied energies are the same as in (4.7) and (4.9). In this case, the new system energy balance is as

$$\mathcal{E}_{stor} + \tilde{\mathcal{E}}_{dis} = \mathcal{E}_{sup}. \qquad (4.16)$$

(4.16) is equal to

$$\mathcal{E}_{stor} - \mathcal{E}_{sup} = -\tilde{\mathcal{E}}_{dis}. \qquad (4.17)$$

Substituting (4.17) into (4.6) leads to

$$\mathcal{H} = \Delta\mathcal{E}_{dis} \qquad (4.18)$$

with $\mathcal{H}$ as in (4.6) and

$$\Delta\mathcal{E}_{dis} = \mathcal{E}_{dis} - \tilde{\mathcal{E}}_{dis} = \int_0^\tau (d(x) - \tilde{d}(x))dt \qquad (4.19)$$

which is the dissipated-energy change.

4.5 Fault detection

In the previous sections, we have studied the energy balance of passive systems and the influence of faults to it. Based on (4.5), (4.12) and (4.18) we have

$$\begin{cases} \mathscr{H} = 0 & \text{for fault-free case} \\ \mathscr{H} = \mathscr{E}_{fault} & \text{for faulty case} \end{cases} \quad (4.20)$$

with $\mathscr{H}$ as in (4.6), and $\mathscr{E}_{fault}$ represents the energy changes:

$$\begin{cases} \mathscr{E}_{fault} = \Delta\mathscr{E}_{stor} & \text{for stored-energy change} \\ \mathscr{E}_{fault} = \Delta\mathscr{E}_{dis} & \text{for dissipated-energy change.} \end{cases}$$

with $\Delta\mathscr{E}_{stor}$ as in (4.13) and $\Delta\mathscr{E}_{dis}$ as in (4.19). Based on (4.20), the fault detection logic can be designed as:

$$\begin{cases} \mathscr{H} \neq 0 \implies \text{faulty} \\ \mathscr{H} = 0 \implies \text{fault-free.} \end{cases} \quad (4.21)$$

In practice, it is more convenient to evaluate $\mathscr{H}$ in a moving time window $[\tau, \tau + \eta]$ as

$$\mathscr{H}(\tau) = \mathscr{E}_{stor}(\tau) + \mathscr{E}_{dis}(\tau) - \mathscr{E}_{sup}(\tau) \quad (4.22)$$

where

$$\begin{aligned} \mathscr{E}_{stor}(\tau) &= V(x(\tau+\eta)) - V(x(\tau)) \\ \mathscr{E}_{dis}(\tau) &= \int_{\tau}^{\tau+\eta} d(x)dt \\ \mathscr{E}_{sup}(\tau) &= \int_{\tau}^{\tau+\eta} y^T u \, dt. \end{aligned} \quad (4.23)$$

In this case, the fault detection logic is that:

$$\begin{cases} \mathscr{H}(\tau) \neq 0 \implies \text{faulty} \\ \mathscr{H}(\tau) = 0 \implies \text{fault-free.} \end{cases} \quad (4.24)$$

4.6 Fault isolation

After the faults have been detected, the next step is to achieve the fault isolation. In Section 4.4, the faults have been defined as different energy changes. Based on it, the first step of the fault isolation is to find out which kind of energy change appears in the system.

4.6.1 Isolation of different energy changes

Firstly, the different properties of energy changes are studied. Based on (4.19), the dissipated-energy change in the moving time window $[\tau,\ \tau+\eta]$ can be written as

$$\Delta\mathscr{E}_{dis}(\tau) = \int_{\tau}^{\tau+\eta} (d(x) - \tilde{d}(x))dt,$$

then according to (4.15) we have

$$\begin{cases} \Delta\mathscr{E}_{dis}(\tau) \leq 0 \text{ for a larger dissipation rate} \\ \Delta\mathscr{E}_{dis}(\tau) \geq 0 \text{ for a smaller dissipation rate} \end{cases}$$

which indicates that, $\Delta\mathscr{E}_{dis}(\tau)$ has the same sign in all evaluation time windows when there is a fault in energy-dissipating components (i.e. a larger or smaller dissipation rate).

Based on (4.13), the stored-energy change in the moving time window $[\tau,\ \tau+\eta]$ is

$$\Delta\mathscr{E}_{stor}(\tau) = \tilde{V}(x(\tau+\eta)) - V(x(\tau+\eta)) + \tilde{V}(x(\tau)) - V(x(\tau)).$$

It can be seen that, the sign of $\Delta\mathscr{E}_{stor}(\tau)$ depends also on system states, which is different from the case of the dissipated-energy change.

In faulty case we have

$$\mathscr{H}(\tau) = \mathscr{E}_{fault}(\tau)$$

where

$$\begin{cases} \mathscr{E}_{fault}(\tau) = \Delta\mathscr{E}_{dis}(\tau) \text{ for dissipated-energy change} \\ \mathscr{E}_{fault}(\tau) = \Delta\mathscr{E}_{stor}(\tau) \text{ for stored-energy change.} \end{cases}$$

So based on the different properties of $\Delta\mathscr{E}_{dis}(\tau)$ and $\Delta\mathscr{E}_{stor}(\tau)$, the type of energy change can be isolated by

$$\begin{cases} \mathscr{H}(\tau) \geq 0\ (\tau \geq \tau_{det})\ \text{ or }\ \mathscr{H}(\tau) \leq 0\ (\tau \geq \tau_{det}) \implies \text{dissipated-energy change} \\ otherwise \implies \text{stored-energy change.} \end{cases} \quad (4.25)$$

Here τ_{det} is the time window when the fault is detected by (4.24). With the knowlege of the type of the energy change, we can distinguish between the faults in energy-storing components and the faults in energy-dissipating components.

4.6.2 Isolation of different components

The second step of energy-balance based fault isolation is to find out which energy-storing or energy-dissipating component is faulty. Since the type of the energy change has been isolated by the first step, only part of the components (energy-storing components or energy-dissipating components) need to be considered. Suppose that, $\mathscr{E}_{fault}(\tau)$ can be written as:

$$\mathscr{E}_{fault}(\tau) = \mathscr{E}_{fault_1}(\tau) + \mathscr{E}_{fault_2}(\tau) + ... + \mathscr{E}_{fault_m}(\tau) \tag{4.26}$$

where $\mathscr{E}_{fault_j}(\tau)(j=1..m)$ is the energy change caused by the fault in j^{th} component,

$$\mathscr{E}_{fault_j}(\tau) = \Delta p_j N_j(\tau), \ j = 1..m, \tag{4.27}$$

and Δp_j is the change of dissipation rate or capacity of jth component. N_i is the function of system states which relates to j^{th} component. It is assumed that, Δp_j is constant. Based on (4.27) we have

$$\frac{\mathscr{E}_{fault_i}(\tau)}{N_j(\tau)} = \Delta p_j, \ when \ N_j(\tau) \neq 0. \tag{4.28}$$

For the case when $N_j(\tau) = 0$,

$$\mathscr{E}_{fault_j}(\tau) = 0. \tag{4.29}$$

Suppose only the kth component is faulty, then

$$\begin{cases} \mathscr{E}_{fault_k}(\tau) \neq 0 \\ \mathscr{E}_{fault_j}(\tau) = 0, \ j = 1..k-1, k+1..m. \end{cases} \tag{4.30}$$

So based on (4.26) we have

$$\mathscr{E}_{fault}(\tau) = \mathscr{E}_{fault_k}(\tau). \tag{4.31}$$

For each component, constructing $M_j(\tau)(j=1..m)$ as

$$M_j(\tau) = \frac{\mathscr{E}_{fault}(\tau)}{N_j(\tau)}, \ when \ N_j(\tau) \neq 0.$$

Based on (4.28) and (4.31), for the kth faulty component, it turns out

$$M_k(\tau) = \Delta p_k. \tag{4.32}$$

Recalling the assumption that Δp_k is constant, $M_k(\tau)$ is also constant, while $M_j(\tau)(j = 1..k-1, k+1..m)$ do not have such property. Based on it and according to (4.29), the faulty component can be isolated by

$$\begin{cases} M_j(\tau) \text{ is constant}, & \text{when } N_j(\tau) \neq 0 \\ \mathscr{E}_{fault}(\tau) = 0, & \text{when } N_j(\tau) = 0 \end{cases} \implies j^{th} \text{ compenent is faulty}$$
$$otherwise \implies j^{th} \text{ compenent is fault-free.} \quad (4.33)$$

Note that $\mathscr{H}(\tau)$ is available as in (4.22), and in faulty case we have

$$\mathscr{H}(\tau) = \mathscr{E}_{fault}(\tau),$$

so in applications, $\mathscr{E}_{fault}(\tau)$ is replaced by $\mathscr{H}(\tau)$ in (4.33).

4.7 Energy-balance based FDI for passive systems with unknown inputs

Since unknown inputs (disturbances) are inevitable in practice, in this section, energy-balance based FDI schemes for passive systems with unknown inputs are studied. The influence of unknown inputs to the energy balance is modeled as unknown-input energy and denoted by $\mathscr{E}_{un}$. Based on (4.4), a new energy balance can be constructed by taking into account $\mathscr{E}_{un}$ as

$$V(x(\tau)) - V(x(0)) + \mathscr{E}_{dis} + \mathscr{E}_{un} = \int_0^\tau y^T u\, dt. \quad (4.34)$$

Evaluated in the time window $[\tau, \tau + \eta]$, energy balance (4.34) can be written as

$$\mathscr{H}(\tau) = \mathscr{E}_{un}(\tau) \quad (4.35)$$

with

$$\mathscr{H}(\tau) = \mathscr{E}_{stor}(\tau) + \mathscr{E}_{dis}(\tau) - \mathscr{E}_{sup}(\tau) \quad (4.36)$$

and $\mathscr{E}_{stor}(\tau)$, $\mathscr{E}_{dis}(\tau)$ and $\mathscr{E}_{sup}(\tau)$ are as in (4.23). $\mathscr{E}_{un}(\tau)$ is the unknown-input energy in the time window $[\tau, \tau + \eta]$, which is assumed to be bounded by

$$|\mathscr{E}_{un}(\tau)| \leq \Theta. \quad (4.37)$$

4.7.1 Fault detection

Following the same definition of the faults as in Section 4.4, we have

$$\begin{cases} \mathscr{H}(\tau) = \mathscr{E}_{un}(\tau) & \text{for fault-free case} \\ \mathscr{H}(\tau) = \mathscr{E}_{fault}(\tau) + \mathscr{E}_{un}(\tau) & \text{for faulty case} \end{cases} \quad (4.38)$$

where

$$\begin{cases} \mathscr{E}_{fault}(\tau) = \Delta\mathscr{E}_{stor}(\tau) & \text{for stored-energy change} \\ \mathscr{E}_{fault}(\tau) = \Delta\mathscr{E}_{dis}(\tau) & \text{for dissipated-energy change.} \end{cases}$$

Then based on the bound of the unknown-input energy $\mathscr{E}_{un}(\tau)$ in (4.37), the fault detection logic is designed as

$$\begin{cases} |\mathscr{H}(\tau)| > \Theta \Longrightarrow \text{faulty} \\ |\mathscr{H}(\tau)| \leq \Theta \Longrightarrow \text{fault-free} \end{cases} \quad (4.39)$$

with $\mathscr{H}(\tau)$ as in (4.36).

4.7.2 Fault isolation

Similar as in Section 4.6, the fault isolation for passive systems with unknown inputs also includes two steps.

Isolation of different energy changes

As discussed in Section 4.6.1, the dissipated-energy change and the stored-energy change have different properties. For dissipated-energy change, $\mathscr{E}_{fault}(\tau)$ has the same sign in all time windows after the fault appears; For stored-energy change, $\mathscr{E}_{fault}(\tau)$ does not have such property. According to (4.38), in the faulty case,

$$\mathscr{H}(\tau) = \mathscr{E}_{fault}(\tau) + \mathscr{E}_{un}(\tau). \quad (4.40)$$

Because of unknown-input energy $\mathscr{E}_{un}(\tau)$, $\mathscr{E}_{fault}(\tau)$ is no more available, so the properties of $\mathscr{E}_{fault}(\tau)$ can not be directly used as in Section 4.6.1. According to (4.39), a fault is detected when $\mathscr{H}(\tau)$ exceeds the range $[-\Theta, \Theta]$ of $\mathscr{E}_{un}(\tau)$. Since $\mathscr{E}_{fault}(\tau)$ will have the same sign when it is dissipated-energy change, in this case, $\mathscr{H}(\tau)$ can only cross one of the bounds Θ or $-\Theta$. For stored-energy change, both Θ and $-\Theta$ could be crossed. Based on this property, assume

Chapter 4 Energy-balance based FDI framework for passive nonlinear systems

τ_{det} is the time window when the fault is detected by (4.39), then the different energy changes can be isolated by

$$\begin{cases} \mathcal{H}(\tau) \geq -\Theta \ (\tau \geq \tau_{det}) \ \text{or} \ \mathcal{H}(\tau) \leq \Theta \ (\tau \geq \tau_{det}) \implies \text{dissipated-energy change} \\ otherwise \implies \text{stored-energy change.} \end{cases} \quad (4.41)$$

Isolation of different components

Since $\mathcal{E}_{fault}(\tau)$ is not available, new isolation method will be developed which is extended from the scheme in Section 4.6.2. Constructing $\mathcal{R}_j(\tau)(j = 1..m)$ as

$$\mathcal{R}_j(\tau) = Nm_{jw}(\tau)\left(\frac{\mathcal{H}(\tau+w)}{N_j(\tau+w)} - \frac{\mathcal{H}(\tau)}{N_j(\tau)}\right),$$
when $N_j(\tau) \neq 0$ and $N_j(\tau+w) \neq 0, \ j = 1..m$ \quad (4.42)

where $N_j(\tau)$ is defined in (4.27), and $Nm_{jw}(\tau)$ is the smaller value of $|N_j(\tau)|$ and $|N_j(\tau+w)|$,

$$Nm_{jw}(\tau) = min\{|N_j(\tau)|, |N_j(\tau+w)|\}.$$

$w > 0$ is a design parameter. We have

$$\begin{aligned}
|\mathcal{R}_j(\tau)| &= Nm_{jw}(\tau)\left|\frac{\mathcal{H}(\tau+w)}{N_j(\tau+w)} - \frac{\mathcal{H}(\tau)}{N_j(\tau)}\right| \\
&= Nm_{jw}(\tau)\left|\frac{\mathcal{E}_{fault}(\tau+w) + \mathcal{E}_{un}(\tau+w)}{N_j(\tau+w)} - \frac{\mathcal{E}_{fault} + \mathcal{E}_{un}(\tau)}{N_j(\tau)}\right| \\
&= Nm_{jw}(\tau)\left|\frac{\mathcal{E}_{fault}(\tau+w)}{N_j(\tau+w)} - \frac{\mathcal{E}_{fault}(\tau)}{N_j(\tau)} + \frac{\mathcal{E}_{un}(\tau+w)}{N_j(\tau+w)} - \frac{\mathcal{E}_{un}(\tau)}{N_j(\tau)}\right| \\
&= Nm_{jw}(\tau)\left|M_j(\tau+w) - M_j(\tau) + \frac{\mathcal{E}_{un}(\tau+w)}{N_j(\tau+w)} - \frac{\mathcal{E}_{un}(\tau)}{N_j(\tau)}\right|
\end{aligned}$$

with $M_j(\tau)$ as in (4.32). For the kth faulty component, as discussed in Section 4.6.2, $M_k(\tau)$ is constant, so that

$$|\mathcal{R}_k(\tau)| = Nm_{kw}(\tau)\left|\frac{\mathcal{E}_{un}(\tau+w)}{N_k(\tau+w)} - \frac{\mathcal{E}_{un}(\tau)}{N_k(\tau)}\right|$$

where $Nm_{kw}(\tau) = min\{|N_k(\tau)|, |N_k(\tau+w)|\}$. Since unknown-input energy is bounded by $\mathcal{E}_{un} \leq \Theta$, we have

$$|\mathcal{R}_k(\tau)| \leq Nm_{kw}(\tau)\frac{2\Theta}{Nm_{kw}(\tau)} = 2\Theta.$$

4.8 RLC Example

For the case when $N_k(\tau) = 0$, we have $\mathscr{E}_{fault}(\tau) = 0$. In this case, the behavior of $\mathscr{H}(\tau)$ is the same as the fault-free case, which fulfills

$$|\mathscr{H}(\tau)| \leq \Theta.$$

So that the isolation scheme can be designed as

$$\begin{cases} |\mathscr{R}_j(\tau)| \leq 2\Theta, \ \tau > \tau_{det} + w \\ \quad \text{when } N_j(\tau) \neq 0 \text{ and } N_j(\tau + w) \neq 0 \implies j^{th} \text{ compenent is faulty} \\ |\mathscr{H}(\tau)| \leq \Theta \ \text{ when } N_j(\tau) = 0 \\ otherwise \quad \implies j^{th} \text{ compenent is fault-free} \end{cases} \quad (4.43)$$

where τ_{det} is the time window when the fault is detected by (4.39). $\mathscr{R}_j(\tau)$ is evaluated when $\tau > \tau_{det} + w$ to avoid the false isolation caused by the case that, $\mathscr{H}(\tau)$ belongs to the fault-free case and $\mathscr{H}(\tau + w)$ belongs to the faulty case.

4.8 RLC Example

In this section, the proposed FDI schemes are illustrated by the RLC example.

4.8.1 System description

Consider the RLC circuit shown in Fig 4.1, which consists of a resistor R_e, an inductor L_d and a capacitor C_a. u is the voltage of the power source, u_c is the voltage of the capacitor and i is the current. With u and i as system input and output, the state space system model of the RLC circuit can be written as

$$\begin{aligned} \dot{x} &= Ax + Bu + d \\ y &= Cx \end{aligned} \quad (4.44)$$

where

$$x = \begin{bmatrix} u_c & i \end{bmatrix}^T, C = \begin{bmatrix} 0 & 1 \end{bmatrix}$$
$$A = \begin{bmatrix} 0 & 1/C_a \\ -1/L_d & -R_e/L_d \end{bmatrix}, B = \begin{bmatrix} 0 \\ 1/L_d \end{bmatrix}$$

and d is the unknown input vector.

4.8.2 Energy balance construction

The storage function of passive system (4.44) can be designed as the energy stored in capacitor and inductor:

$$V = \frac{1}{2}C_a u_c^2 + \frac{1}{2}L_d i^2 = \frac{1}{2}x^T P x \qquad (4.45)$$

with

$$P = \begin{bmatrix} C_a & 0 \\ 0 & L_d \end{bmatrix}. \qquad (4.46)$$

In order to obtain the system energy balance, deriving the storage function V along the trajectory of system (4.44) we get

$$\begin{aligned}\frac{\partial V}{\partial x}\dot{x} &= x^T P(Ax + Bu + d) \\ &= \frac{1}{2}x^T(PA + A^T P)x + x^T PBu + x^T Pd.\end{aligned} \qquad (4.47)$$

Noticing that

$$PA + A^T P = \begin{bmatrix} 0 & 0 \\ 0 & -2R_e \end{bmatrix}, \quad PB = \begin{bmatrix} 0 & 1 \end{bmatrix}^T = C^T,$$

(4.47) can be written as

$$\frac{\partial V}{\partial x}\dot{x} = -i^2 R_e + iu + x^T Pd. \qquad (4.48)$$

Integrating (4.48) from 0 to τ leads to

$$V(x(\tau)) - V(x(0)) + \int_0^\tau i^2 R_e dt - \int_0^\tau iu dt = \int_0^\tau (x^T Pd) dt. \qquad (4.49)$$

It holds

$$\int_0^\tau i^2 R_e dt \geq 0$$

which is the dissipated energy according to the definition in (4.3). From a physic point of view, $\int_0^\tau i^2 R_e dt$ is the energy dissipated by the resistor. $\int_0^\tau iu dt$ in (4.49) is the supplied energy which is the electrical power flowing into the RLC circuit.

(4.49) can be rewritten as the following energy balance:

$$\mathcal{H} = \mathcal{E}_{un} \tag{4.50}$$

where

$$\begin{aligned}
\mathcal{H} &= \mathcal{E}_{stor} + \mathcal{E}_{dis} - \mathcal{E}_{sup}, \\
\mathcal{E}_{stor} &= V(x(\tau)) - V(x(0)), \\
\mathcal{E}_{dis} &= \int_0^\tau i^2 R_e dt, \\
\mathcal{E}_{sup} &= \int_0^\tau iu\, dt, \\
\mathcal{E}_{un} &= \int_0^\tau (x^T Pd)dt.
\end{aligned}$$

4.8.3 Fault detection

Evaluated in the time window $[\tau,\ \tau+\eta]$, the unknown-input energy is

$$\mathcal{E}_{un}(\tau) = \int_\tau^{\tau+\eta} (x^T Pd)dt. \tag{4.51}$$

Substituting (4.46) into (4.51) we have

$$\mathcal{E}_{un}(\tau) = \int_\tau^{\tau+\eta} (P_{11}x_1 d_1 + P_{22}x_2 d_2)dt \tag{4.52}$$

where $x_i(i=1,2)$ and $d_i(i=1,2)$ are i^{th} element of vector x and d. Suppose the system states and unknown inputs are bounded by

$$\begin{aligned}
x_1^2 &\leq \delta_{x1}^2,\ x_2^2 \leq \delta_{x2}^2 \\
d_1^2 &\leq \delta_{d1}^2,\ d_2^2 \leq \delta_{d2}^2,
\end{aligned}$$

it turns out

$$|\mathcal{E}_{un}(\tau)| \leq \Theta$$

with

$$\Theta = \eta P_{11}\delta_{x1}\delta_{d1} + \eta P_{22}\delta_{x2}\delta_{d2} \tag{4.53}$$

and P_{jj} being the jth diagonal element of matrix P. The fault detection logic can then be designed as:

$$\begin{cases} |\mathcal{H}(\tau)| > \Theta \implies \text{faulty} \\ |\mathcal{H}(\tau)| \leq \Theta \implies \text{fault-free} \end{cases} \quad (4.54)$$

where

$$\begin{aligned} \mathcal{H}(\tau) &= \mathcal{E}_{sup}(\tau) + \mathcal{E}_{dis}(\tau) - \mathcal{E}_{stor}(\tau), \\ \mathcal{E}_{stor}(\tau) &= V(x(\tau+\eta)) - V(x(\tau)), \\ \mathcal{E}_{dis}(\tau) &= \int_{\tau}^{\tau+\eta} i^2 R_e dt, \\ \mathcal{E}_{sup}(\tau) &= \int_{\tau}^{\tau+\eta} iudt. \end{aligned} \quad (4.55)$$

4.8.4 Fault isolation

Isolation of different energy changes

The faults in the capacitor or inductor will lead to stored-energy change, and faults in the resistor will lead to dissipated-energy change. These two kinds of energy changes can be isolated by logic (4.25).

Isolation of different components

Since there are two energy-storing components, isolation logic should be designed to distinguish the faults between them. Following the proceedure in Section 4.6.2, the stored-energy change should be first divided into energy changes of different componenents. Based on the storage function in (4.45), the stored-energy change in the time window $[\tau, \ \tau + \eta]$ can be written as

$$\mathcal{E}_{fault}(\tau) = \mathcal{E}_{fault_1}(\tau) + \mathcal{E}_{fault_2}(\tau)$$

where

$$\mathcal{E}_{fault_1}(\tau) = \Delta C_a N_1(\tau), \ \mathcal{E}_{fault_2}(\tau) = \Delta L_d N_2(\tau)$$

and

$$N_1(\tau) = \frac{1}{2}x(\tau+\eta)^T q_1 x(\tau+\eta) - \frac{1}{2}x(\tau)^T q_1 x(\tau),$$
$$N_2(\tau) = \frac{1}{2}x(\tau+\eta)^T q_2 x(\tau+\eta) - \frac{1}{2}x(\tau)^T q_2 x(\tau),$$
$$q_1 = \begin{bmatrix} 1 & 0 \\ 0 & 0 \end{bmatrix}, \quad q_2 = \begin{bmatrix} 0 & 0 \\ 0 & 1 \end{bmatrix}.$$

$\mathscr{E}_{fault_1}(\tau)$ and $\mathscr{E}_{fault_2}(\tau)$ are the energy changes caused by the fault in capacitor and inductor. ΔC_a and ΔL_d are the changes of the capacitance and inductance. Constructing $\mathscr{R}_i(\tau)(i=1,2)$ as in (4.42), faults in the capacitor and inductor can then be isolated by logic (4.43).

4.8.5 Parameter setting and threshold computation

In the simulation study, the parameters of system (4.44) are set as

$$R_e = 10\Omega, \ L_d = 20H, \ C_a = 0.5F$$

and the unknown input vector is as

$$d = \begin{bmatrix} d_1 & d_2 \end{bmatrix}^T = \begin{bmatrix} 0 & 0.02\cos(0.002t) \end{bmatrix}^T.$$

So the bounds of the unknown inputs are

$$d_1^2 \le \delta_{d1}^2 = 0, \quad d_2^2 \le \delta_{d2}^2 = 0.02^2.$$

The following input signal is considered:

$$u = 60\sin^2(0.1t) + 30\cos(0.1t).$$

Driven by input signal u, the bounds of the system states in fault-free case are

$$x_1^2 \le \delta_{x1}^2 = 35^2, \quad x_2^2 \le \delta_{x2}^2 = 4^2.$$

The width of the evaluation time window and the isolation window is set as $\eta = 0.1s$ and $w = 1s$. Then according to (4.53), threshold Θ is as

$$\Theta = \eta P_{11}\delta_{x1}\delta_{d1} + \eta P_{22}\delta_{x2}\delta_{d2} = 0.16.$$

Chapter 4 Energy-balance based FDI framework for passive nonlinear systems

For numerical reasons, $\mathcal{R}_j(\tau)(j = 1, 2)$ are calculated when $|N_j(\tau)| > 0.001$ and $|N_j(\tau + w)| > 0.001$. The isolation logic (4.43) becomes

$$\begin{cases} |\mathcal{R}_j(\tau)| \leq 2\Theta, & \tau > \tau_{det} + w \\ |\mathcal{H}(\tau)| \leq \Theta & \text{when } |N_j(\tau)| \leq 0.001 \end{cases} \implies j^{th} \text{ compenent is faulty}$$

$$otherwise \implies j^{th} \text{ compenent is fault-free.} \quad (4.56)$$

For the case when $|N_j(\tau)| \leq 0.001$ or $|N_j(\tau + w)| \leq 0.001$, $\mathcal{R}_j(\tau)$ is set to be zero.

4.8.6 Simulation results

For different kinds of faults, simulations of 400 seconds have been carried out. In all simulations, the fault appears at t=200 seconds. In the figures, the solid line represents the behavior of $\mathcal{H}(\tau)$ or $\mathcal{R}_j(\tau)(j = 1, 2)$, and the dashed line represents the threshold Θ and $-\Theta$.

Fault-free case: Fig 4.2 shows the behavior of $\mathcal{H}(\tau)$ in fault-free case. Due to the unknown-input energy, $\mathcal{H}(\tau)$ is not equal to zero. Since it is always fulfilled that $|\mathcal{H}(\tau)| \leq \Theta$, there are no false alarms.

Fault in resistor: Fig 4.3 shows the simulation results, when the resistor is faulty. Two abrupt faults which lead to changes of the resistance are considered. For the first fault, the change of resistance is $\Delta R_e = 5\Omega$, and for the second fault $\Delta R_e = -2\Omega$. It can be observed from the figures that, after the faults appear, it turns to be $|\mathcal{H}(\tau)| > \Theta$, which leads to a successful fault detection. On the other hand, in both situations, $\mathcal{H}(\tau)$ crosses only one of the thresholds, so based on the isolation logic (4.25), the dissipated-energy change is successfully isolated.

Fault in capacitor: Fig 4.4 shows the simulation results, when the capacitor is faulty. An abrupt fault which leads to the change $\Delta C_a = -0.1F$ of the capacitance is considered. As shown in Fig 4.4a, it turns to be $|\mathcal{H}(\tau)| > \Theta$ after the fault appears which leads to a successful fault detection. On the other hand, since $\mathcal{H}(\tau)$ crosses both upper and lower thresholds, the stored-energy change is isolated according to the isolation logic (4.25). In order to further isolate the faulty energy-storing component, the behavior of $\mathcal{R}_1(\tau)$ and $\mathcal{R}_2(\tau)$ are shown in Fig 4.4b and Fig 4.4c. For $\tau > \tau_{det} + w$ ($\tau_{det} = 200.2s, w = 1s$), only $\mathcal{R}_1(\tau)$ stays between the thresholds ($\mathcal{R}_1(\tau)$ crosses the threshold only when $\tau < \tau_{det} + w$), since $|\mathcal{H}(\tau)| \leq \Theta$ is fulfilled when $|M_{Q_1}(\tau)| \leq 0.001$ as shown in Fig 4.4d, according to isolation logic (4.56), a fault in capacitor is successfully isolated.

Fault in inductor: Fig 4.5 shows the results, when the inductor is faulty. An abrupt fault which leads to the change $\Delta L_d = -5H$ of the inductance is considered. From Fig 4.5a we can see that, $\mathscr{H}(\tau)$ crosses both thresholds, so stored-energy change is isolated. It can be observed from Fig 4.5b and Fig 4.5c that, only $\mathscr{R}_2(\tau)$ stays between the thresholds for $\tau > \tau_{det} + w$. Together with the behavior of $|\mathscr{H}(\tau)| \leq \Theta$ when $|M_{Q_2}(\tau)| \leq 0.001$ in Fig 4.5d, according to isolation logic (4.56), a fault in inductor is successfully isolated.

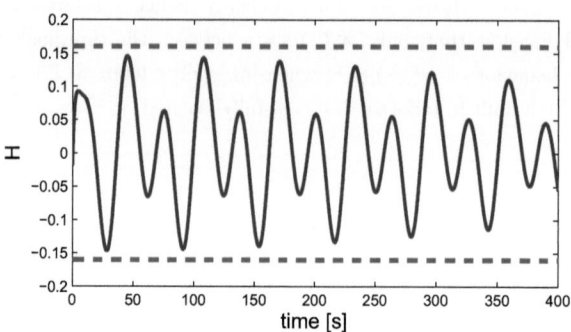

Figure 4.2: Behavior of $\mathcal{H}(\tau)$ in fault-free case

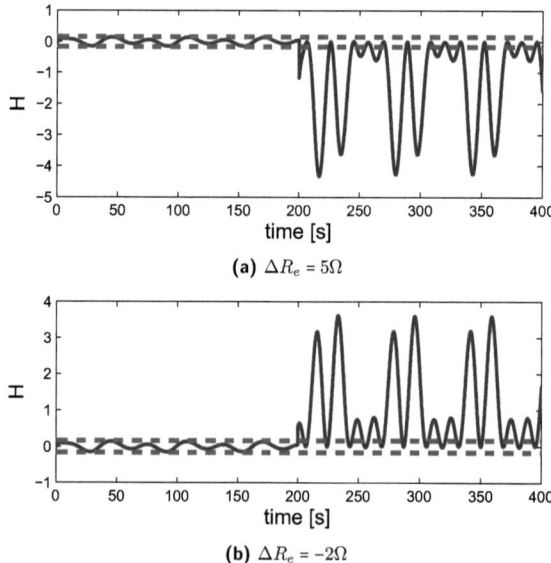

(a) $\Delta R_e = 5\Omega$

(b) $\Delta R_e = -2\Omega$

Figure 4.3: Faults in resistor. Figure (a) shows the behavior of $\mathcal{H}(\tau)$ when an abrupt resistance change $\Delta R_e = 5\Omega$ occurs at $t = 200$ seconds. Figure (b) shows the result when the abrupt resistance change is $\Delta R_e = -2\Omega$.

4.8 RLC Example

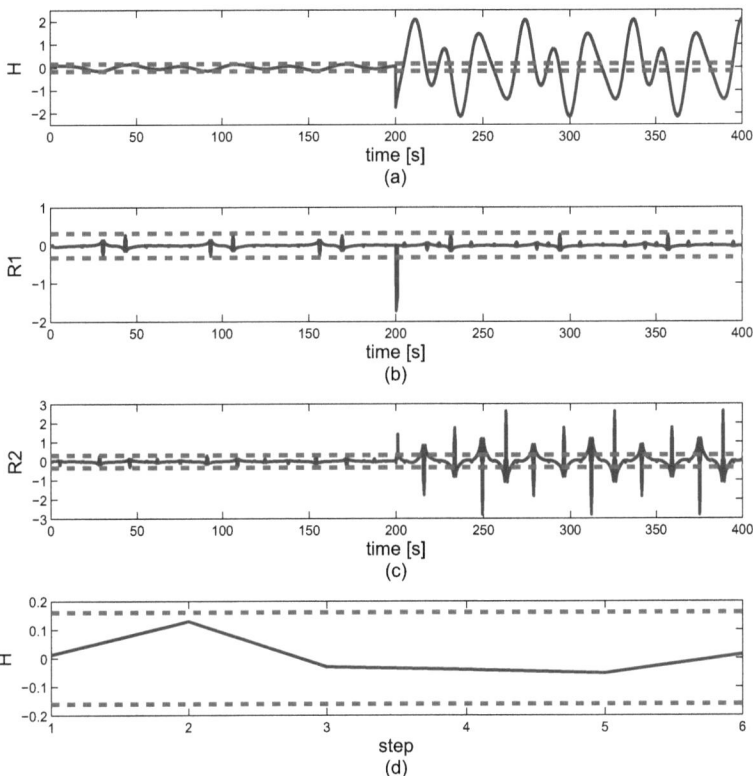

Figure 4.4: Fault in capacitor. An abrupt fault which leads to capacitance change $\Delta C_a = -0.1 F$ occurs at $t = 200$ seconds. Figure (a) shows the behavior of $\mathcal{H}(\tau)$. Figure (b),(c) show the behavior of $\mathcal{R}_1(\tau)$ and $\mathcal{R}_2(\tau)$. Figure (d) shows the behavior of $\mathcal{H}(\tau)$ when $|M_1(\tau)| \leq 0.001$.

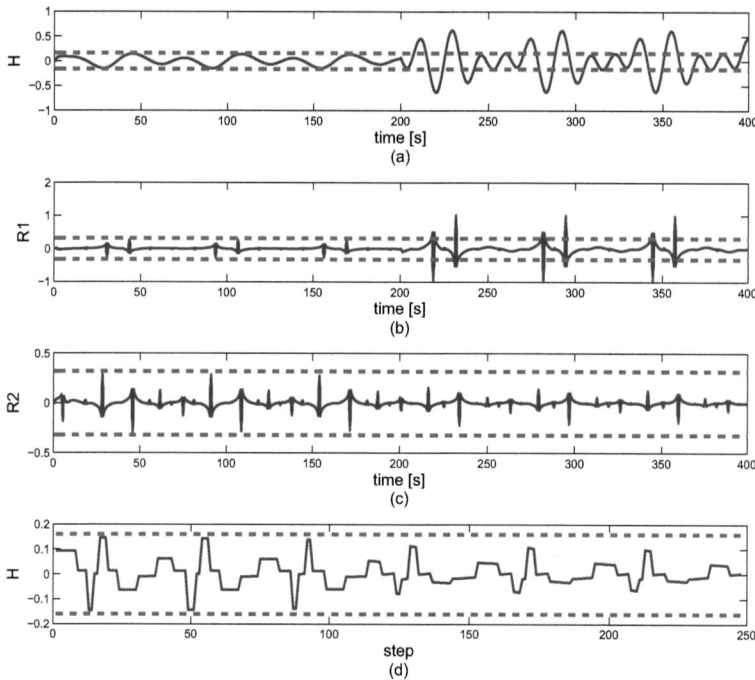

Figure 4.5: Fault in inductor. An abrupt fault which leads to inductance change $\Delta L_d = -5H$ occurs at $t = 200$ seconds. Figure (a) shows the behavior of $\mathcal{H}(\tau)$. Figure (b),(c) show the behavior of $\mathcal{R}_1(\tau)$ and $\mathcal{R}_2(\tau)$. Figure (d) shows the behavior of $\mathcal{H}(\tau)$ when $|M_2(\tau)| \leq 0.001$.

4.9 Summary

This chapter has proposed a complete energy-balance based FDI framework for nonlinear passive systems. Based on passivity, an energy balance which includes stored, dissipated and supplied energies has been established. Faults have been defined according to their different influences on system energies. Fault detection has been achieved by checking the validity of the energy balance. For fault isolation, a two-step approach has been proposed, which firstly isolates the class of the fault and secondly isolates the faulty component. The thresholds were designed based on the bound of unknown-input energy. The effectiveness of the proposed FDI schemes has been illustrated by an RLC circuit example.

Chapter 5

Design of Energy-balance based FDI for two classes of passive nonlinear systems

The objective of this chapter is to propose design procedures of energy-balance based FDI schemes for two classes of passive nonlinear systems: (1) Input-affine passive systems; (2) Lagrangian systems. The design procedures consist of finding a storage function, establishing the energy balance and computation of the threshold. The proposed design procedures are also applied to linear passive systems, which can be considered as a special case of input-affine passive systems.

5.1 Input-affine passive systems

5.1.1 Introduction

The class of input-affine systems is one of the most important nonlinear models that has been intensivly studied since decades [22]. It has the following formulation:

$$\begin{aligned} \dot{x} &= f(x) + g(x)u \\ y &= h(x) \end{aligned} \quad (5.1)$$

where $x \in \Re^n$ is the state vector, $u \in \Re^m$ is the control input and $y \in \Re^m$ is the output vector. In the passive theory, system (5.1) is also widely used since many results for linear passive systems can be extended to this kind of nonlinear systems [76]. The passivity of system (5.1) can be checked by the following Lemma.

5.1 Input-affine passive systems

Lemma 5.1.1. *[71] System* (5.1) *is passive, if there exists a nonnegative function* $V(x)$, *with* $V(x(0)) = 0$, *such that*

$$\frac{\partial V}{\partial x} f(x) \leq 0 \tag{5.2}$$

$$\frac{\partial V}{\partial x} g(x) = h^T(x). \tag{5.3}$$

Lemma 5.1.1 can be considered as an extention of the Kalman-Yakubovich-Popov Lemma to the nonlinear case.

5.1.2 Energy balance of input-affine passive systems

Suppose that we have already found the nonnegative function V which fulfills the conditions (5.2) and (5.3) in Lemma 5.1.1, then V can be used as the storage function of system (5.1). The derivative of V along the trajectory of the system (5.1) is

$$\frac{dV}{dt} = \frac{\partial V}{\partial x} f(x) + \frac{\partial V}{\partial x} g(x) u. \tag{5.4}$$

Integrating (5.4) from 0 to τ leads to

$$V(x(\tau)) - V(x(0)) = \int_0^\tau \frac{\partial V}{\partial x} f(x) dt + \int_0^\tau \frac{\partial V}{\partial x} g(x) u dt. \tag{5.5}$$

According to (5.3) we have

$$\int_0^\tau \frac{\partial V}{\partial x} g(x) u dt = \int_0^\tau h^T(x) u dt = \int_0^\tau y^T u dt, \tag{5.6}$$

and by substituting (5.6) into (5.5) we get

$$V(x(\tau)) - V(x(0)) - \int_0^\tau \frac{\partial V}{\partial x} f(x) dt = \int_0^\tau y^T u dt. \tag{5.7}$$

(5.7) is an energy balance of passive nonlinear system (5.1). Based on (5.2), it holds that

$$-\int_0^\tau \frac{\partial V}{\partial x} f(x) dt \geq 0 \tag{5.8}$$

which represents the dissipated energy. The passivity of system (5.1) can be proved by substituting (5.8) into (5.7), which leads to

$$V(x(\tau)) - V(x(0)) \leq \int_0^\tau y^T u dt.$$

Chapter 5 Design of Energy-balance based FDI for two classes of passive nonlinear systems

Assuming the zero initial condition (i.e. $V(x(0)) = 0$), energy balance (5.7) can be rewritten as

$$\mathscr{H} = 0 \tag{5.9}$$

where

$$\mathscr{H} = \mathscr{E}_{stor} + \mathscr{E}_{dis} - \mathscr{E}_{sup} \tag{5.10}$$

and the stored, dissipated and supplied energies are as follows:

$$\mathscr{E}_{stor} = V(x(\tau)) \tag{5.11}$$

$$\mathscr{E}_{dis} = -\int_0^\tau \frac{\partial V}{\partial x} f(x) dt \tag{5.12}$$

$$\mathscr{E}_{sup} = \int_0^\tau y^T u \, dt. \tag{5.13}$$

Based on energy balance (5.9), fault detection scheme can then be designed following Chapter 4. The major task is to find the storage function V, which can be achieved by solving nonlinear partial differential inequalities (5.2) and (5.3) [77, 78]. An alternative way is to analyze the stored energy of the system physically. For fault isolation, energy changes caused by the fault in different components need to be studied, which depend on the structure of the considered system.

5.1.3 A design example

Consider a pendulum shown in Fig 5.1. The length of the rod is denoted as l and the mass of the bob is denoted as m. The rod is assumed to be rigid and has zero mass. θ denotes the angle subtended by the rod and the vertical axis through the pivot point. The pendulum is free to swing in the vertical plane and is driven by a torque T_q. The friction force is assumed to be proportional to the speed of the bob with a friction coefficient k. The state space model of the pendulum can be written as

$$\begin{aligned} \dot{x} &= f(x) + q(x)u \\ y &= h(x) \end{aligned} \tag{5.14}$$

with

$$x = \begin{bmatrix} \theta & \dot{\theta} \end{bmatrix}^T, \; f(x) = \begin{bmatrix} -\frac{g}{l}\sin\theta - \frac{k}{m}\dot{\theta} & \dot{\theta} \end{bmatrix}^T$$
$$q(x) = \begin{bmatrix} \frac{1}{ml^2} & 0 \end{bmatrix}^T, \; u = T_q, \; h(x) = \dot{\theta}$$

5.1 Input-affine passive systems

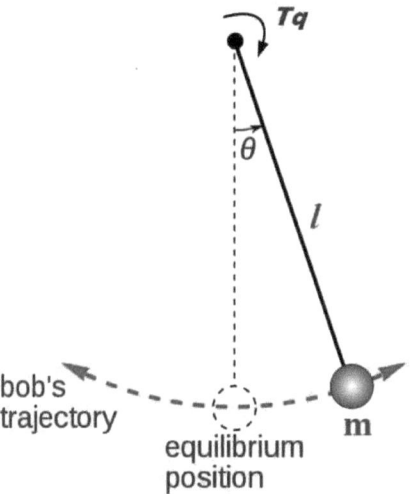

Figure 5.1: Pendulum

and g is the acceleration due to gravity. The stored energy in the pendulum system includes the kinetic energy $\mathscr{E}_k$ and the potential energy $\mathscr{E}_p$ of the bob,

$$\begin{aligned}\mathscr{E}_k &= \frac{1}{2}ml^2\dot{\theta}^2, \\ \mathscr{E}_p &= l(1-\cos\theta)mg,\end{aligned}$$

so a candidate for the storage function can be designed as

$$V = \frac{1}{2}ml^2\dot{\theta}^2 + l(1-\cos\theta)mg.$$

As a result, we have

$$\begin{aligned}\frac{\partial V}{\partial x}f(x) &= ml^2\dot{\theta}(-\frac{g}{l}\sin\theta - \frac{k}{m}\dot{\theta}) + lmg\dot{\theta}\sin\theta \\ &= -kl^2\dot{\theta}^2 \\ &\leq 0, \\ \frac{\partial V}{\partial x}g(x) &= ml^2\dot{\theta}\frac{1}{ml^2} \\ &= \dot{\theta} \\ &= h^T(x),\end{aligned}$$

so that conditions (5.2) and (5.3) in Lemma 5.1.1 are fulfilled. According to (5.9), the energy balance of the system (5.14) can be constructed as

$$\mathscr{H} = \mathscr{E}_{stor} + \mathscr{E}_{dis} - \mathscr{E}_{sup} = 0 \qquad (5.15)$$

where the stored, dissipated and supplied energies are as:

$$\mathscr{E}_{stor} = \frac{1}{2}ml^2\dot{\theta}^2 + l(1-cos\theta)mg \qquad (5.16)$$

$$\mathscr{E}_{dis} = \int_0^\tau kl^2\dot{\theta}^2 dt \qquad (5.17)$$

$$\mathscr{E}_{sup} = \int_0^\tau T_q\dot{\theta} dt. \qquad (5.18)$$

Two kinds of faults are considered. The first kind of fault is the change of the friction coefficient k, which will change the dissipated energy in (5.17). This energy change can be written as

$$\Delta\mathscr{E}_{fault_1} = \int_0^\tau \Delta kl^2\dot{\theta}^2 dt \qquad (5.19)$$

with Δk as the variation of the friction coefficient. The second kind of fault is the change of the mass m of the bob, which can be for example caused by an unknown object adhered to the bob. This kind of fault will lead to the change of stored energy in (5.16) as

$$\Delta\mathscr{E}_{fault_2} = \frac{1}{2}\Delta ml^2\dot{\theta}^2 + l(1-cos\theta)\Delta mg \qquad (5.20)$$

with Δm as the variation of the mass. Based on the energy balance (5.15) and energy changes (5.19) and (5.20), the fault detection and isolation schemes can then be designed following Chapter 4.

5.1.4 Application to linear passive systems

For linear passive systems, the energy balance and possible energy changes can be obtained in a systematical way. The following linear system is considered:

$$\begin{aligned} \dot{x} &= Ax + Bu \\ y &= Cx \end{aligned} \qquad (5.21)$$

5.1 Input-affine passive systems

where $x \in \mathfrak{R}^n$ is the state vector, $u \in \mathfrak{R}^m$ is the input vector and $y \in \mathfrak{R}^m$ is the output vector. For this kind of system, the storage function can be designed as

$$V(x) = \frac{1}{2} x^T P x, \tag{5.22}$$

where P is a real symmetric matrix which fulfills $P > 0$. (5.22) has been widely used in Lyapunov stability analysis [79]. For system (5.21), based on storage function $V(x)$, passive conditions (5.2) and (5.3) becomes:

$$PA + A^T P \leq 0 \tag{5.23}$$

$$PB = C^T. \tag{5.24}$$

(5.23) and (5.24) are linear inequalities, so P can be obtained by powerful LMI tools. According to (5.9), the energy balance for system (5.21) is constructed as

$$\mathscr{H} = \mathscr{E}_{stor} + \mathscr{E}_{dis} - \mathscr{E}_{sup} = 0 \tag{5.25}$$

with

$$\mathscr{E}_{stor} = \frac{1}{2} x^T P x \tag{5.26}$$

$$\mathscr{E}_{dis} = -\frac{1}{2} \int_0^\tau x^T (PA + A^T P) x \, dt \tag{5.27}$$

$$\mathscr{E}_{sup} = \int_0^\tau y^T u \, dt. \tag{5.28}$$

Based on (5.25), fault detection scheme can be designed. For fault isolation purpose, the possible energy changes are analyzed in the following.

First consider the stored energy changes. Since P is a real symmetric matrix, there exists an orthogonal matrix Q_s such that

$$Q_s^T P Q_s = \Lambda_s$$

with Λ_s as a diagonal matrix whose entries are the eigenvalues of P. If we define a state transformation as $x = Q_s \tilde{x}$, then based on the new states $\tilde{x}$, the stored energy $\mathscr{E}_{stor}$ in (5.26) becomes

$$\mathscr{E}_{stor} = \frac{1}{2} \tilde{x}^T \Lambda_s \tilde{x}. \tag{5.29}$$

63

Because Λ_s is a diagonal matrix, $\mathscr{E}_{stor}$ can be rewritten as

$$\mathscr{E}_{stor} = \sum_{j=1}^{n} \mathscr{E}_{stor_j} \qquad (5.30)$$

where

$$\mathscr{E}_{stor_j} = \frac{1}{2}\Lambda_{s_j} M_{s_j}(\tilde{x}), \quad M_{s_j}(\tilde{x}) = \tilde{x}_j^2$$

and Λ_{s_j} is the jth entry of Λ_s. (5.30) indicates that, the stored energy $\mathscr{E}_{stor}$ can be divided into n parts $\mathscr{E}_{stor_j}(j = 1..n)$, each of them can be considered as the energy stored in one "component". Here different energy-storing "components" means their relations to system states (i.e. $M_{s_j}(\tilde{x})$) are different. $\Lambda_{s_j}(j = 1..n)$ represents the energy storage ability of different energy-storing components. Faults can lead to change of $\Lambda_{s_j}(j = 1..n)$ which results in stored-energy change as

$$\Delta\mathscr{E}_{fault_j} = \frac{1}{2}\Delta\Lambda_{s_j} M_{s_j}(\tilde{x}), \quad M_{s_j}(x) = \tilde{x}_j^2, \; j = 1..n \qquad (5.31)$$

Secondly, consider the dissipated energy changes. Let

$$P_A = -(PA + A^T P).$$

P_A is also a real symmetric matrix, similarly there exists an orthogonal matrix Q_d such that

$$Q_d^T P_A Q_d = \Lambda_d$$

with Λ_d as a diagonal matrix whose entries are the eigenvalues of P_A. Defining a state transformation as $x = Q_d \bar{x}$, then based on the new states $\bar{x}$, the dissipated energy $\mathscr{E}_{dis}$ in (5.27) can be rewritten as

$$\mathscr{E}_{dis} = \sum_{j=1}^{n} \mathscr{E}_{dis_j}$$

where

$$\mathscr{E}_{dis_j} = \frac{1}{2}\int_0^\tau \Lambda_{d_j} M_{d_j}(\bar{x}) dt, \quad M_{d_j}(\bar{x}) = \bar{x}_j^2$$

and Λ_{d_j} is the jth entry of Λ_d. Since P_A may have eigenvalues equal to zero, some entries Λ_{d_j} could also be zero. Without loss of generality, suppose that, there are $\bar{n}$ non-zero entries $\Lambda_{d_j}(j = 1..\bar{n})$, then the dissipated energy $\mathscr{E}_{dis}$ can be divided into $\bar{n}$ parts, each of them can

5.1 Input-affine passive systems

be considered as the energy dissipated by one "component". $\Lambda_{d_j}(j = 1..\bar{n})$ represents the energy dissipation rate of different energy-dissipating components. Faults can lead to change of $\Lambda_{d_j}(j = 1..\bar{n})$ which results in dissipated-energy change as

$$\Delta \mathscr{E}_{fault_j} = \frac{1}{2} \int_0^\tau \Delta \Lambda_{d_j} M_{d_j}(\bar{x}) dt, \quad M_{d_j}(x) = \bar{x}_i^2, \quad j = 1..\bar{n}. \tag{5.32}$$

Based on (5.31) and (5.32), faults in different energy-storing and energy-dissipating components can then be isolated following Chapter 4.

5.1.5 Input-affine passive systems with unknown inputs

When there are unknown inputs, system (5.1) becomes

$$\dot{x} = f(x) + g(x)u + d$$
$$y = h(x).$$

where d is the unknown input vector. It is assumed that

$$d^T d \leq \delta_d^2. \tag{5.33}$$

Following a similar way as in Section 5.1.2, we can get the new energy balance as

$$\mathscr{H} = \mathscr{E}_{un} \tag{5.34}$$

with $\mathscr{H}$ as in (5.10) and the unknown-input energy is

$$\mathscr{E}_{un} = \int_0^\tau (\frac{\partial V}{\partial x} d) dt.$$

Evaluated in the time window $[\tau, \tau + \eta]$, (5.34) becomes

$$\mathscr{H}(\tau) = \mathscr{E}_{un}(\tau) \tag{5.35}$$

where

$$\begin{aligned}
\mathscr{H}(\tau) &= \mathscr{E}_{stor}(\tau) + \mathscr{E}_{dis}(\tau) - \mathscr{E}_{sup}(\tau) \\
\mathscr{E}_{stor}(\tau) &= V(x(\tau+\eta)) - V(x(\tau)) \\
\mathscr{E}_{dis}(\tau) &= -\int_{\tau}^{\tau+\eta} \frac{\partial V}{\partial x} f(x) dt \\
\mathscr{E}_{sup}(\tau) &= \int_{\tau}^{\tau+\eta} y^T u\, dt \\
\mathscr{E}_{un}(\tau) &= \int_{\tau}^{\tau+\eta} (\frac{\partial V}{\partial x} d) dt.
\end{aligned} \qquad (5.36)$$

Assume that

$$(\frac{\partial V}{\partial x})^T \frac{\partial V}{\partial x} \leq \delta_v^2,$$

together with (5.33) we have

$$|\mathscr{E}_{un}(\tau)| \leq \Theta, \quad \Theta = \eta \delta_d \delta_v. \qquad (5.37)$$

Based on the energy balance (5.35) and threshold Θ, the fault detection scheme is designed as

$$\begin{cases} |\mathscr{H}(\tau)| > \Theta \implies \text{faulty} \\ |\mathscr{H}(\tau)| \leq \Theta \implies \text{fault-free} \end{cases}$$

with $\mathscr{H}(\tau)$ as in (5.36).

The design procedure of the energy-balance based FDI for input-affine passive systems is summarized as:

1. Finding the storage function V which fulfills (5.2) and (5.3).

2. Establishing the energy balance as in (5.35) and computing the threshold Θ as in (5.37) to achieve the fault detection.

3. Analyzing the possible energy changes and designing the isolation logic following Chapter 4.

5.2 Lagrangian systems

5.2.1 Introduction

Lagrangian systems arise from variational calculus and gave a first general definition of physical dynamical systems in analytical mechanics [76, 80]. Compared with the classical mechanics which is based on vector forces and velocities, analytical mechanics uses two scalar properties i.e. the kinetic and potential energies to analyze the mechanical motion. The application of Lagrangian systems is not only limited to mechanical systems, it can also be used to describe the dynamics of various engineering systems as electrical circuits or electro-mechanical systems. It is also widely used to derive different control laws by taking into account the structure of the system dynamics derived from energy based modeling [81]. Since Lagrangian systems relate closely to system energies, they have great potential for the application of energy-balance based FDI schemes.

Definition 5.2.1. *[76] The Lagrangian (L_a) of a dynamic system is a function that summarizes the dynamics of the system.*

Mathmatically, consider a configuration manifold $Q = \mathfrak{R}^n$, the points on this manifold are denoted by $q \in \mathfrak{R}^n$ and are called generalized coordinates. Denote by $TQ = \mathfrak{R}^{2n}$ its tangent bundle and its elements by $(q, \dot{q}) \in \mathfrak{R}^{2n}$ where $\dot{q}$ is called generalized velocity, then the Lagrangian is defined by a real function $L_a(q, \dot{q})$ [76]. If the Lagrangian of a system is known, then the equations of the system dynamics can be obtained by a direct substitution of the expression for the Lagrangian into the *Euler-Lagrange* equation, which is defined as

$$\frac{d}{dt}\left(\frac{\partial L_a}{\partial \dot{q}}(q, \dot{q})\right) - \frac{\partial L_a}{\partial q}(q, \dot{q}) = 0. \tag{5.38}$$

The concept of Lagrangian was originally introduced in a reformulation of classical mechanics by Irish mathematician *William Rowan Hamilton* known as Lagrangian mechanics. For mechanical systems, the Lagrangian is defined as the kinetic energy $\mathscr{E}_k$ of the system minus its potential energy $\mathscr{E}_p$. In symbols,

$$L_a = \mathscr{E}_k - \mathscr{E}_p.$$

Example: Let us consider a simple example of the linear mass-spring system consisting of a mass (m) attached to a fixed frame through a spring. The position of the mass with respect to the fixed frame is denoted as q and the elasticity coefficient of the spring is denoted as r_e. The

Chapter 5 Design of Energy-balance based FDI for two classes of passive nonlinear systems

Lagrangian function is given by ([76])

$$L_a(q, \dot{q}) = \mathscr{E}_k - \mathscr{E}_p, \quad \mathscr{E}_k = \frac{1}{2}m\dot{q}^2, \quad \mathscr{E}_p = \frac{1}{2}r_e q^2 \tag{5.39}$$

where $\mathscr{E}_k$ is the kinetic energy of the mass and $\mathscr{E}_p$ is the potential energy of the spring. Subsituting the Lagrangian function (5.39) into Euler-Lagrange equation (5.38) we have

$$m\ddot{q} + kq = 0$$

which is exactly the dynamics equation of the linear mass-spring system.

5.2.2 Lagrangian systems with external forces and dissipation

Defining $u \in \mathfrak{R}^n$ as the vector of generalized forces acting on the system, and considering *Rayleigh dissipation function* $R(\dot{q})$ which satisfies [76]

$$\dot{q}^T \frac{\partial R}{\partial \dot{q}}(\dot{q}) \geq 0, \tag{5.40}$$

then based on Euler-Lagrange equation (5.38), Lagrangian systems with external forces (input) u and dissipation can be written as

$$\frac{d}{dt}(\frac{\partial L_a}{\partial \dot{q}}(q, \dot{q})) - \frac{\partial L_a}{\partial q}(q, \dot{q}) + \frac{\partial R}{\partial \dot{q}} = u. \tag{5.41}$$

For the linear mass-spring example, when friction is considered, the dissipation function is as

$$R(\dot{q}) = \frac{1}{2}k\dot{q}^2 \tag{5.42}$$

with k as the friction coefficient. Substituting (5.42) into (5.41), it turns out

$$m\ddot{q} + r_e q + k\dot{q} = u$$

which is the dynamics equation of the linear mass-spring system with external force and dissipation.

5.2.3 Stored energy of Lagrangian systems

As discussed in Section 5.2.1, Lagrangian L_a relates closely to the energies stored in the system. Based on $L_a(q, \dot{q})$, the storage function V can be obtained by the *Legendre transformation* with respect to the generalized velocity $\dot{q}$ as [76]

$$V(q, \dot{q}) = \dot{q}^T p - L_a(q, \dot{q}) \tag{5.43}$$

where p is the vector of generalized momenta,

$$p = \frac{\partial L_a(q,\dot{q})}{\partial \dot{q}}.$$

For the linear mass-spring system, based on the Lagrangian (5.39) we have

$$\begin{aligned} V(q,\dot{q}) &= \dot{q}^T m\dot{q} - (\frac{1}{2}m\dot{q}^2 - \frac{1}{2}r_e q^2) \\ &= \frac{1}{2}m\dot{q}^2 + \frac{1}{2}r_e q^2 \\ &= \mathscr{E}_k + \mathscr{E}_p. \end{aligned}$$

It can be seen that, the obtained storage function $V(q,\dot{q})$ is exactly the sum of the stored energies including kinetic and potential energies.

5.2.4 Passivity of Lagrangian systems

Lemma 5.2.1. *Lagrangian system (5.41) with input u and output $\dot{q}$ is passive with respect to storage function (5.43).*

Proof. The derivative of the storage function $V(q,\dot{q})$ along the trajectory of system (5.41) is

$$\begin{aligned} \frac{dV(q,\dot{q})}{dt} &= \ddot{q}^T \frac{\partial L_a}{\partial \dot{q}} + \dot{q}^T \frac{d}{dt}p - (\ddot{q}^T \frac{\partial L_a}{\partial \dot{q}} + \dot{q}^T \frac{\partial L_a}{\partial q}) \\ &= \dot{q}^T \frac{d}{dt}p - \dot{q}^T \frac{\partial L_a}{\partial q} \\ &= \dot{q}^T (\frac{d}{dt}(\frac{\partial L_a}{\partial \dot{q}}(q,\dot{q})) - \frac{\partial L_a}{\partial q}(q,\dot{q})). \end{aligned} \qquad (5.44)$$

Based on system equation (5.41) we have

$$\frac{d}{dt}(\frac{\partial L_a}{\partial \dot{q}}(q,\dot{q})) - \frac{\partial L_a}{\partial q}(q,\dot{q}) = u - \frac{\partial R}{\partial \dot{q}}. \qquad (5.45)$$

Substituting (5.45) into (5.44) yields

$$\begin{aligned} \frac{dV(q,\dot{q})}{dt} &= \dot{q}^T(u - \frac{\partial R}{\partial \dot{q}}) \\ &= \dot{q}^T u - \dot{q}^T \frac{\partial R}{\partial \dot{q}}. \end{aligned} \qquad (5.46)$$

Integrating (5.46) from 0 to τ leads to

$$V(\tau) - V(0) = \int_0^\tau \dot{q}^T u \, dt - \int_0^\tau \dot{q}^T \frac{\partial R}{\partial \dot{q}} dt \qquad (5.47)$$

which is the energy balance of system (5.41). According to (5.40), it holds

$$\int_0^\tau \dot{q}^T \frac{\partial R}{\partial \dot{q}} dt \geq 0,$$

so we have

$$V(\tau) - V(0) \leq \int_0^\tau \dot{q}^T u \, dt$$

which indicates that system (5.41) is passive. $\square$

5.2.5 Energy balance of Lagrangian systems

Assuming the zero initial condition (i.e. $V(0) = 0$), the energy balance (5.47) of system (5.41) can be written as

$$\mathcal{H} = 0 \tag{5.48}$$

where

$$\mathcal{H} = \mathcal{E}_{stor} + \mathcal{E}_{dis} - \mathcal{E}_{sup} \tag{5.49}$$

and the stored, dissipated and supplied energies are respectively as follows:

$$\begin{aligned}
\mathcal{E}_{stor} &= V(\tau) - V(0) \\
\mathcal{E}_{dis} &= \int_0^\tau \dot{q}^T \frac{\partial R}{\partial \dot{q}} dt \\
\mathcal{E}_{sup} &= \int_0^\tau \dot{q}^T u \, dt.
\end{aligned} \tag{5.50}$$

5.2.6 Lagrangian systems with unknown inputs

When there are unknown inputs, Lagrangian system (5.41) becomes

$$\frac{d}{dt}\left(\frac{\partial L_a}{\partial \dot{q}}(q, \dot{q})\right) - \frac{\partial L_a}{\partial q}(q, \dot{q}) + \frac{\partial R}{\partial \dot{q}} + d = u. \tag{5.51}$$

where d is the unknown input vector. It is assumed that

$$d^T d \leq \delta_d^2. \tag{5.52}$$

5.2 Lagrangian systems

Following a similar procedure as in the proof of Lemma 5.2.1, we can get the energy balance of system (5.51) as

$$\mathcal{H} = \mathcal{E}_{un} \tag{5.53}$$

with $\mathcal{H}$ as in (5.49) and the unknown-input energy is

$$\mathcal{E}_{un} = -\int_0^\tau (\dot{q}^T d) dt. \tag{5.54}$$

Evaluated in the time window $[\tau, \tau + \eta]$, (5.53) becomes

$$\mathcal{H}(\tau) = \mathcal{E}_{un}(\tau) \tag{5.55}$$

where

$$\begin{aligned}
\mathcal{H}(\tau) &= \mathcal{E}_{stor}(\tau) + \mathcal{E}_{dis}(\tau) - \mathcal{E}_{sup}(\tau) \\
\mathcal{E}_{stor}(\tau) &= V(\tau + \eta) - V(\tau) \\
\mathcal{E}_{dis}(\tau) &= \int_\tau^{\tau+\eta} \frac{\partial R}{\partial \dot{q}} dt \\
\mathcal{E}_{sup}(\tau) &= \int_\tau^{\tau+\eta} \dot{q}^T u \, dt \\
\mathcal{E}_{un}(\tau) &= -\int_\tau^{\tau+\eta} (\dot{q}^T d) dt.
\end{aligned} \tag{5.56}$$

Assume that

$$\dot{q}^T \dot{q} \le \delta_{\dot{q}}^2,$$

together with (5.52) we have

$$|\mathcal{E}_{un}(\tau)| \le \Theta, \quad \Theta = \eta \delta_d \delta_{\dot{q}}. \tag{5.57}$$

Based on the energy balance (5.55) and threshold Θ, the fault detection scheme is designed as

$$\begin{cases} |\mathcal{H}(\tau)| > \Theta \implies \text{faulty} \\ |\mathcal{H}(\tau)| \le \Theta \implies \text{fault-free} \end{cases}$$

with $\mathcal{H}(\tau)$ as in (5.56). For fault isolation purpose, since Lagrangian systems have very clear physical meanings, the functions of possible energy changes can be obtained by physically analyzing the systems.

The design procedure of energy-balance based FDI for Lagrangian systems is summarized as:

1. Constructing the storage function by (5.43) based on the Lagrangian $L_a(q,\dot{q})$ of the system.

2. Establishing the energy balance as in (5.56) and computing the threshold Θ as in (5.57) to achieve the fault detection.

3. Analyzing the possible energy changes and designing the isolation logic following Chapter 4.

This design procedure will also be illustrated by a benchmark study in the next chapter.

5.3 Summary

In this chapter, the design procedures of energy-balance based FDI have been proposed for input-affine passive systems and Lagrangian systems, which consist of finding a storage function, establishing the energy balance and computation of the threshold. The proposed design procedure has also been applied to linear passive systems, which can be considered as a special case of input-affine passive systems.

Chapter 6

Application to the robot manipulator benchmark

The objective of this chapter is to apply the energy-balance based FDI schemes to robot manipulator benchmark, which has strong nonlinearities and is widely used in factory automation systems. Since in practice robot manipulators are usually driven by DC motors, the dynamics of DC motors is also included in the benchmark model. This electro-mechanical system can be modeled as an interconnection of two Lagrangian systems, which is an excellent benchmark for illustrating the usefulness of the proposed energy-balance based FDI framework.

6.1 Introduction

The FDI problem of robot manipulator has received considerable attentions, since it is widely used for factory automation systems and currently also employed in scenarios requiring high degree of autonomy (e.g. space and underwater missions, rescue operations etc.) [82]. Various approaches have been proposed. Due to the strong nonlinearities in manipulators, simplified model has been used in [83]. The full nonlinear model has been applied in [84, 85] for the observer-based methods and in [86] for the filtering methods, in which the nonlinearities has been compensated using the full state measurement. High gain observer and sliding mode observer based approaches have been developed in [87] and [88]. The stability of these two kinds of observers have been proved in an attracting region of system states. In this chapter, the proposed energy-balance based FDI schemes are applied to robot manipulator. Compared with the exsiting methods, the required on-line computation is much lower, and the difficulties in stabilizing the nonlinear observers like in [87, 88] have also been released. On the other hand, robot manipulators are usually driven by DC motors, and the effect of neglecting the dynamics of DC motors may deteriorate the control and FDI system performances [89], so in the benchmark study, an electro-mechanical system which includes both dynamics is considered.

6.2 System model and description of faults

The dynamics of robot manipulator driven by armature-controlled DC motors can be considered as the interconnection of manipulator subsystem and DC motor subsystem. Both of them can be modeled as Lagrangian systems [76]. For robot manipulator subsystem, the Lagrangian function is given by

$$L_a(q,\dot{q}) = T(q,\dot{q}) - U(q) \tag{6.1}$$

with q as the vector of link angles of manipulator and

$$T(q,\dot{q}) = \frac{1}{2}\dot{q}^T M(q)\dot{q} \tag{6.2}$$

is the kinetic energy of the manipulator. Matrix $M(q)$ is positive definite and is called the inertia matrix. $U(q)$ is a real function which represents the potential energy of the manipulator. The energy dissipation is caused by friction which has the following dissipation function:

$$R(\dot{q}) = \frac{1}{2}\dot{q}^T K \dot{q} \tag{6.3}$$

where K is a diagonal matrix whose entries are friction rates of the links of manipulator. The input u of the robot manipulator system is the mechanical torque generated by DC motors,

$$u = W_t i_c \tag{6.4}$$

where W_t is a diagonal matrix whose entries are the torque constants of DC motors, and i_c is the vector of armature currents. Recall the model of the Lagrangian system with unknown inputs derived in Chapter 5:

$$\frac{d}{dt}\left(\frac{\partial L_a}{\partial \dot{q}}(q,\dot{q})\right) - \frac{\partial L_a}{\partial q}(q,\dot{q}) + \frac{\partial R}{\partial \dot{q}} + d = u. \tag{6.5}$$

Substituting (6.1), (6.3) and (6.4) into (6.5) we have the model of manipulator subsystem as

$$M(q)\ddot{q} + C(q,\dot{q})\dot{q} + g(q) + K\dot{q} + d_m = W_t i_c \tag{6.6}$$

with

$$g(q) = \frac{dU}{dq}(q), \quad C(q,\dot{q}) = \sum_{k=1}^{n_m} \Gamma_{ijk}\dot{q}_k.$$

Γ_{ijk} are called *Christoffel's symbols*, its detailed description can be found in [90]. n_m is the number of the links of manipulator. $C(q,\dot{q})$ has an important property that $\dot{M}(q) - 2C(q,\dot{q})$ is skew-symmetric which leads to ([76])

$$\dot{q}^T(\dot{M}(q) - 2C(q,\dot{q}))\dot{q} = 0. \tag{6.7}$$

6.2 System model and description of faults

d_m is the unknown input vector of manipulator subsystem. It is assumed that

$$d_m^T d_m \leq \delta_m^2. \tag{6.8}$$

For DC motor subsystem, the Lagrangian function is given by

$$L_a(\tilde{q}, \dot{\tilde{q}}) = \frac{1}{2}\dot{\tilde{q}}^T Q_{in} \dot{\tilde{q}}, \quad \dot{\tilde{q}} = i_c. \tag{6.9}$$

Here Q_{in} is a diagonal matrix whose entries representing the inductances of DC motors. The energy dissipation of DC motor subsystem is caused by resistance. The dissipation function is as

$$R(\dot{\tilde{q}}) = \frac{1}{2}\dot{\tilde{q}}^T R_e \dot{\tilde{q}} = \frac{1}{2}i_c^T R_e i_c \tag{6.10}$$

where $R_e \in \mathfrak{R}^{n_d \times n_d}$ is a diagonal matrix whose entries are the resistances of DC motors, and n_d is the number of DC motors. There are two kinds of inputs in DC motor subsystem, the first is the armature voltage u_v, and the second is the voltage generated by the rotating of the rotors as

$$u_r = -W_t \dot{q},$$

so the total input for DC motor subsystem is

$$u = u_v + u_r = u_v - W_t \dot{q}. \tag{6.11}$$

Substituting (6.9), (6.10) and (6.11) into (6.5) we have the model of DC motor subsystem as

$$Q_{in}\dot{i}_c + R_e i_c + d_c = u_v - W_t \dot{q} \tag{6.12}$$

where d_c is the unknown input vector of the DC motor subsystem with elements $d_{cj}(j = 1..n_d)$. It is assumed that

$$d_c^T d_c \leq \delta_c^2, \quad d_{cj}^2 \leq \delta_{cj}^2, \quad j = 1..n_d. \tag{6.13}$$

Subsystems (6.6) and (6.12) are coupled with each other, the argumented system can be constructed as

$$\begin{aligned} M(q)\ddot{q} + C(q,\dot{q})\dot{q} + g(q) + K\dot{q} + d_m &= W_t i_c \\ Q_{in}\dot{i}_c + R_e i_c + d_c + W_t \dot{q} &= u_v \end{aligned} \tag{6.14}$$

Two kinds of faults are considered in the benchmark study:

75

Dissipated-energy change. As discussed above, system energy will be dissipated by friction in manipulator and resistance in DC motors. Typically, mechanical faults in manipulator will lead to a lager friction rate, which will increase the dissipated energy. In DC motors, faults could increase or decrease the resistance, which leads to the corresponding change of dissipated energy.

Stored-energy change. The stored energies in system are kinetic and potential energy of manipulator, and the energy stored in the inductor of DC motors. The stored energy of manipulator will be changed when the physical properties (weight, inertia...) are changed by faults, which could be for example caused by an unexpected load on manipulator. The stored energy of DC motors will be changed when the inductance is changed by faults.

6.3 Energy-based FDI system design

6.3.1 Energy balances construction

Recall the formulation of stored energy of Lagrangian system derived in Chapter 5:

$$V(q,\dot{q}) = \dot{q}^T \frac{\partial L_a(q,\dot{q})}{\partial \dot{q}} - L_a(q,\dot{q}). \tag{6.15}$$

Substituting the Lagrangian function (6.1) into (6.15) results in the stored energy of manipulator subsystem

$$\mathcal{E}_{Mstor} = \frac{1}{2}\dot{q}^T M(q)\dot{q} + U(q) = T(q,\dot{q}) + U(q) \tag{6.16}$$

which is the sum of the kinetic energy and the potential energy of the manipulator. Similarly, substituting (6.3), (6.4) and unknown input d_m into (5.50) and (5.54) we have the dissipated, supplied and unknown-input energies of manipulator subsystem as

$$\mathcal{E}_{Mdis} = \int_0^\tau \dot{q}^T K \dot{q} dt$$

$$\mathcal{E}_{Msup} = \int_0^\tau \dot{q}^T W_t i_c dt$$

$$\mathcal{E}_{Mun} = -\int_0^\tau \dot{q}^T d_m dt. \tag{6.17}$$

6.3 Energy-based FDI system design

From a physical point of view, $\mathscr{E}_{Mdis}$ is the energy dissipated by the friction and $\mathscr{E}_{Msup}$ is the energy supplied by the DC motors. The energy balance of manipulator subsystem can then be constructed as

$$\mathscr{H}_M = \mathscr{E}_{Mun} \tag{6.18}$$

with

$$\mathscr{H}_M = \mathscr{E}_{Mstor} + \mathscr{E}_{Mdis} - \mathscr{E}_{Msup}. \tag{6.19}$$

Similarly, for DC motor subsystem, substituting the Lagrangian function (6.9) into (6.15) we get the stored energy as

$$\mathscr{E}_{Dstor} = \frac{1}{2} i_c^T Q_{in} i_c \tag{6.20}$$

which is the energy stored in the inductor of DC motors. The dissipated, supplied and uncertian energy can be obtained by substituting (6.10), (6.11) and unknown input d_c into (5.50) and (5.54):

$$\begin{aligned}
\mathscr{E}_{Ddis} &= \int_0^\tau i_c^T R_e i_c dt \\
\mathscr{E}_{Dsup} &= \int_0^\tau (u_v - W_t \dot{q})^T i_c dt \\
\mathscr{E}_{Dun} &= -\int_0^\tau i_c^T d_c dt.
\end{aligned} \tag{6.21}$$

From a physical point of view, $\mathscr{E}_{Ddis}$ is the energy dissipated by the resistance of DC motors, and $\mathscr{E}_{Dsup}$ is the supplied energy from the electric power source minus the energy flowing to the manipulator subsystem. The energy balance of DC motor subsystem is then constructed as

$$\mathscr{H}_D = \mathscr{E}_{Dun} \tag{6.22}$$

with

$$\mathscr{H}_D = \mathscr{E}_{Dstor} + \mathscr{E}_{Ddis} - \mathscr{E}_{Dsup}. \tag{6.23}$$

Based on energy balances (6.18) and (6.22), fault detection can be designed separately for two subsystems. An alternative approach is to construct an energy balance for the argumented system (6.14) which contains both subsystems, in this way, the detection algorithm can be simplified. From a physical point of view, the stored (dissipated, unknown-input) energy of the

Chapter 6 Application to the robot manipulator benchmark

argumented system (6.14) should be the sum of the stored (dissipated, unknown-input) energy of the two subsystems:

$$\begin{aligned}
\mathscr{E}_{stor} &= \mathscr{E}_{Mstor} + \mathscr{E}_{Dstor} = \frac{1}{2}\dot{q}^T M(q)\dot{q} + U(q) + \frac{1}{2}i_c^T Q_{in} i_c \\
\mathscr{E}_{dis} &= \mathscr{E}_{Mdis} + \mathscr{E}_{Ddis} = \int_0^\tau \dot{q}^T K\dot{q}\,dt + \int_0^\tau i_c^T R_e i_c\,dt \\
\mathscr{E}_{un} &= \mathscr{E}_{Mun} + \mathscr{E}_{Dun} = -\int_0^\tau \dot{q}^T d_m - \int_0^\tau i_c^T d_c\,dt
\end{aligned} \quad (6.24)$$

and the supplied energy of the argumented system is the energy from the electric power source:

$$\mathscr{E}_{sup} = \int_0^\tau u_v^T i_c\,dt. \quad (6.25)$$

Since $\int_0^\tau \dot{q}^T W_t i_c\,dt$ in (6.17) and $-\int_0^\tau \dot{q}^T W_t i_c\,dt$ in (6.21) are the energy exchange between two subsystems, they do not exist in the supplied energy of the argumented system. The energy balance of argumented system (6.14) is as

$$\mathscr{H} = \mathscr{E}_{un} \quad (6.26)$$

with

$$\mathscr{H} = \mathscr{E}_{stor} + \mathscr{E}_{dis} - \mathscr{E}_{sup}. \quad (6.27)$$

Energy balance (6.26) has been established from a physical point of view, it is proved mathematically in the following.

The stored energy function $\mathscr{E}_{stor}$ of the argumented system has been obtained in (6.24), the derivative of $\mathscr{E}_{stor}$ along the trajectory of argumented system (6.14) is

$$\begin{aligned}
\dot{\mathscr{E}}_{stor} &= \dot{q}^T M(q)\ddot{q} + \frac{1}{2}\dot{q}^T \dot{M}(q)\dot{q} + g(q)\dot{q} + i_c^T Q_{in}\dot{i}_c \\
&= \dot{q}^T[W_t i_c - C(q,\dot{q})\dot{q} - g(q) - K\dot{q} - d_m] + \frac{1}{2}\dot{q}^T \dot{M}(q)\dot{q} + g(q)\dot{q} + i_c^T Q_{in}\dot{i}_c \\
&= \dot{q}^T W_t i_c + \frac{1}{2}\dot{q}^T[\dot{M}(q) - 2C(q,\dot{q})]\dot{q} - \dot{q}^T K\dot{q} - \dot{q}^T d_m + i_c^T Q_{in}\dot{i}_c.
\end{aligned}$$

With property (6.7), it turns out

$$\dot{\mathscr{E}}_{stor} = \dot{q}^T W_t i_c - \dot{q}^T K\dot{q} - \dot{q}^T d_m + i_c^T L_a \dot{i}_c. \quad (6.28)$$

According to system model (6.14) we have

$$W_t \dot{q} = u_v - R_e i_c - Q_{in}\dot{i}_c - d_c,$$

6.3 Energy-based FDI system design

and because W_t, R_e and Q_{in} are diagonal matrice, it can be transformed to

$$\dot{q}^T W_t = u_v^T - i_c^T R_e - \dot{i}_c^T Q_{in} - d_c^T. \tag{6.29}$$

Substituting (6.29) into (6.28) yields

$$\begin{aligned}\dot{\mathscr{E}}_{stor} &= [u_v^T - i_c^T R_e - \dot{i}_c^T L_a - d_c^T]i_c - \dot{q}^T K \dot{q} - \dot{q}^T d_m + i_c^T L_a \dot{i}_c \\ &= u_v^T i_c - i_c^T R_e i_c - d_c^T i_c - \dot{q}^T K \dot{q} - \dot{q}^T d_m.\end{aligned} \tag{6.30}$$

Integrating (6.30) from 0 to τ leads to energy balance (6.26).

6.3.2 Fault detection

The fault detection scheme for the whole benchmark system (6.14) is designed based on the energy balance (6.26). Evaluated in a time window $[\tau,\ \tau + \eta]$, the unknown-input energy becomes

$$\mathscr{E}_{un}(\tau) = -\int_\tau^{\tau+\eta} (\dot{q}^T d_m + i_c^T d_c)dt. \tag{6.31}$$

Suppose in fault-free case $\dot{q}$ and i_c are bounded by

$$\dot{q}^T \dot{q} \le \delta_{\dot{q}}^2,\ \ i_c^T i_c \le \delta_i^2,\ \ i_{cj}^2 \le \delta_{ij}^2,\ j = 1..n_d \tag{6.32}$$

where i_{cj} is the jth element of i_c. Then together with the bounds of the unknown inputs in (6.8) and (6.13) we have

$$|\mathscr{E}_{un}(\tau)| \le \Theta \tag{6.33}$$

with

$$\Theta = \eta(\delta_{\dot{q}} \delta_m + \delta_i \delta_c). \tag{6.34}$$

The fault detection logic is designed as

$$\begin{cases} |\mathscr{H}(\tau)| > \Theta \implies \text{faulty} \\ |\mathscr{H}(\tau)| \le \Theta \implies \text{fault-free} \end{cases} \tag{6.35}$$

where

$$\mathcal{H}(\tau) = \mathcal{E}_{stor}(\tau) + \mathcal{E}_{dis}(\tau) - \mathcal{E}_{sup}(\tau)$$
$$\mathcal{E}_{stor}(\tau) = \frac{1}{2}\dot{q}(\tau+\eta)^T M(q(\tau+\eta))\dot{q}(\tau+\eta) + U_g(q(\tau+\eta)) + \frac{1}{2}i_c(\tau+\eta)^T Q_{in} i_c(\tau+\eta)$$
$$\quad - [\frac{1}{2}\dot{q}(\tau)^T M(q(\tau))\dot{q}(\tau) + U_g(q(\tau)) + \frac{1}{2}i_c(\tau)^T Q_{in} i_c(\tau)]$$
$$\mathcal{E}_{dis}(\tau) = \int_\tau^{\tau+\eta} \dot{q}^T K \dot{q} dt + \int_\tau^{\tau+\eta} i_c^T R_e i_c dt$$
$$\mathcal{E}_{sup}(\tau) = \int_\tau^{\tau+\eta} u_v^T i_c dt.$$

6.3.3 Fault isolation

Based on the energy balances (6.18) and (6.22), the faults can be first distinguished between two subsystems. After that, following the schemes proposed in Chapter 4, the faults in each subsystem can be further isolated.

Manipulator subsystem. Evaluated in the time window $[\tau, \tau+\eta]$, the unknown-input energy of the manipulator subsystem becomes

$$\mathcal{E}_{Mun}(\tau) = -\int_\tau^{\tau+\eta} \dot{q}^T d_m dt. \tag{6.36}$$

According to (6.8) and (6.32) we have

$$|\mathcal{E}_{Mun}(\tau)| \leq \Theta_M \tag{6.37}$$

with

$$\Theta_M = \eta \delta_{\dot{q}} \delta_m. \tag{6.38}$$

The faults in the manipulator subsystem can be detected by

$$\begin{cases} |\mathcal{H}_M(\tau)| > \Theta_M \implies \text{manipulator subsystem is faulty} \\ |\mathcal{H}_M(\tau)| \leq \Theta_M \implies \text{manipulator subsystem is fault-free} \end{cases} \tag{6.39}$$

where

$$\mathcal{H}_M(\tau) = \mathcal{E}_{Mstor}(\tau) + \mathcal{E}_{Mdis}(\tau) - \mathcal{E}_{Msup}(\tau)$$

$$\mathcal{E}_{Mstor}(\tau) = \frac{1}{2}\dot{q}(\tau+\eta)^T M(q(\tau+\eta))\dot{q}(\tau+\eta) + U_g(q(\tau+\eta))$$

$$- [\frac{1}{2}\dot{q}(\tau)^T M(q(\tau))\dot{q}(\tau) + U_g(q(\tau))]$$

$$\mathcal{E}_{Mdis}(\tau) = \int_{\tau}^{\tau+\eta} \dot{q}^T K \dot{q} dt$$

$$\mathcal{E}_{Msup}(\tau) = \int_{\tau}^{\tau+\eta} \dot{q}^T W_t i_c dt.$$

The second step of fault isolation is to find out which kind of fault appears in manipulator subsystem, i.e. whether it is dissipated-energy change caused by the change in friction rate or stored-energy change caused by the change in physical properties of manipulator. These two kinds of energy change can be isolated following the proposed scheme in Chapter 4 as follows:

$$\begin{cases} \mathcal{H}_M(\tau) \geq -\Theta_M \ (\tau \geq \tau_{det}) \ \ or \ \ \mathcal{H}_M(\tau) \leq \Theta_M \ (\tau \geq \tau_{det}) \\ \implies \text{dissipated-energy change in manipulator subsystem} \\ otherwise \ \ \implies \text{stored-energy change in manipulator subsystem} \end{cases} \quad (6.40)$$

where τ_{det} is the time window when the fault is detected by (6.35).

The last step is to isolate the faulty link of the manipulator subsystem. Since multi-link manipulator ($n_m > 1$) are considered, when dissipated-energy change is isolated, it is necessary to find out in which link the friction rate is changed. According to (6.17), the dissipated energy of manipulator subsystem in the time window $[\tau, \ \tau+\eta]$ is

$$\mathcal{E}_{Mdis}(\tau) = \int_{\tau}^{\tau+\eta} \dot{q}^T K \dot{q} dt,$$

and because K is a diagonal matrix, it can be rewritten as

$$\mathcal{E}_{Mdis}(\tau) = \sum_{j=1}^{n_m} K_j \int_{\tau}^{\tau+\eta} \dot{q}_j^2 dt \quad (6.41)$$

with $K_j(j=1..n_m)$ as entries of diagonal matrix K. Each of $K_j(j=1..n_m)$ is the friction rate of one link. The energy change caused by the change of friction rate is as

$$\Delta \mathcal{E}_{Mdis}(\tau) = \sum_{j=1}^{n_m} \Delta K_j \int_{\tau}^{\tau+\eta} \dot{q}_j^2 dt \quad (6.42)$$

where $\Delta K_j (j = 1..n_m)$ are the changes of the friction rates. Following Chapter 4, constructing $\mathscr{R}_{Mj}(\tau)(j = 1..n_m)$ as

$$\mathscr{R}_{Mj}(\tau) = Nm_{jw}(\tau)\big(\frac{\mathscr{H}_M(\tau+w)}{N_j(\tau+w)} - \frac{\mathscr{H}_M(\tau)}{N_j(\tau)}\big), \quad N_j(\tau) = \int_\tau^{\tau+\eta} \dot{q}_j^2 dt$$
$$\text{when } N_j(\tau) \neq 0 \text{ and } N_j(\tau+w) \neq 0, \ i = 1..n_m \tag{6.43}$$

with $M_{jw}(\tau)$ as the smaller value of $N_j(\tau)$ and $N_j(\tau+w)$,

$$Nm_{jw}(\tau) = min\{|N_j(\tau)|, |N_j(\tau+w)|\}.$$

and $w > 0$ is a design parameter. For numerical reason, $\mathscr{R}_{Mj}(\tau)(j = 1..n_m)$ are calculated when $|N_j(\tau)| > 0.001$ and $|N_j(\tau+w)| > 0.001$, and they are set to be zero when $|N_i(\tau)| \leq 0.001$ or $|N_i(\tau+w)| \leq 0.001$. The faulty link can be isolated by

$$\begin{cases} |\mathscr{R}_{Mj}(\tau)| \leq 2\Theta_M, \ \tau > \tau_{det} + w \\ |\mathscr{H}_M(\tau)| \leq \Theta_M, \ \text{when } |N_i(\tau)| \leq 0.001 \end{cases}$$
$$\implies j^{th} \text{ link of manipulator is faulty}$$
$$otherwise \implies j^{th} \text{ link of manipulator is fault-free} \tag{6.44}$$

where τ_{det} is the time window when the fault is detected by (6.35).

DC motor subsystem. According to (6.13) and (6.32), the unknown-input energy of the DC motor subsystem in time window $[\tau, \tau+\eta]$ is bounded by

$$|\mathscr{E}_{Dun}(\tau)| \leq \Theta_D, \quad \Theta_D = \eta \delta_i \delta_c \tag{6.45}$$

The faults in DC motor subsystem can be detected by

$$\begin{cases} |\mathscr{H}_D(\tau)| > \Theta_D \implies \text{DC motor subsystem is faulty} \\ |\mathscr{H}_D(\tau)| \leq \Theta_D \implies \text{DC motor subsystem is fault-free} \end{cases} \tag{6.46}$$

where

$$\begin{aligned}
\mathscr{H}_D(\tau) &= \mathscr{E}_{Dstor}(\tau) + \mathscr{E}_{Ddis}(\tau) - \mathscr{E}_{Dsup}(\tau) \\
\mathscr{E}_{Dstor}(\tau) &= \frac{1}{2} i_c(\tau+\eta)^T Q_{in} i_c(\tau+\eta) \\
&\quad -\frac{1}{2} i_c(\tau)^T Q_{in} i_c(\tau) \\
\mathscr{E}_{Ddis}(\tau) &= \int_\tau^{\tau+\eta} i_c^T R_e i_c dt \\
\mathscr{E}_{Dsup}(\tau) &= \int_\tau^{\tau+\eta} (u_v - W_t \dot{q})^T i_c dt.
\end{aligned}$$

6.3 Energy-based FDI system design

The second step of the fault isolation for DC motor subsystem is to find out which motor is faulty. Since Q_{in}, R_e and W_t are diagonal matrice, the jth entry of them (Q_{inj}, R_{ej} and W_{tj}) represents the inductance and resisitance of jth motor, and the jth element i_{cj} of vector i_c also represents the armature current in j^{th} motor. So that, energy balance (6.22) of DC motor subsystem can be divided to energy balances for each motor. Based on it, faults in j^{th} motor can be isolated by

$$\begin{cases} |\mathcal{H}_{Dj}(\tau)| > \Theta_{Dj} \implies j^{th} \text{ DC motor is faulty} \\ |\mathcal{H}_{Dj}(\tau)| \leq \Theta_{Dj} \implies j^{th} \text{ DC motor is fault-free} \end{cases} \quad (6.47)$$

where

$$\begin{aligned}
\mathcal{H}_{Dj}(\tau) &= \mathcal{E}_{Dstorj}(\tau) + \mathcal{E}_{Ddisj}(\tau) - \mathcal{E}_{Dsupj}(\tau) \\
\mathcal{E}_{Dstorj}(\tau) &= \frac{1}{2} i_{cj}(\tau+\eta)^T Q_{inj} i_{cj}(\tau+\eta) \\
&\quad - \frac{1}{2} i_{cj}(\tau)^T Q_{inj} i_{cj}(\tau) \\
\mathcal{E}_{Ddisj}(\tau) &= \int_{\tau}^{\tau+\eta} i_{cj}^T R_{ej} i_{cj} dt \\
\mathcal{E}_{Dsupj}(\tau) &= \int_{\tau}^{\tau+\eta} (u_{vj} - W_{tj} \dot{q}_j)^T i_{cj} dt \\
\Theta_{Dj} &= \eta \delta_{ij} \delta_{cj}.
\end{aligned}$$

The last step is to find out which kind of fault appears in the faulty DC motor, i.e. whether it is dissipated-energy change caused by the change of resistance or stored-energy change caused by the change of inductance. Suppose j^{th} motor is faulty, following the isolation logic (4.41) in Chapter 4, two different kinds of faults can be isolated by

$$\begin{cases} \mathcal{H}_{Dj}(\tau) \geq -\Theta_{Dj} \; (\tau \geq \tau_{det}) \; \text{ or } \; \mathcal{H}_{Dj}(\tau) \leq \Theta_{Dj} \; (\tau \geq \tau_{det}) \\ \implies \text{resistance change (dissipated-energy change) in } j^{th} \text{ motor} \\ otherwise \quad \implies \text{inductance change (stored-energy change) in } j^{th} \text{ motor} \end{cases} \quad (6.48)$$

with τ_{det} as the time window when the fault is detected by (6.35). Since there is only one energy-dissipating component (resistance) and one energy-storing component (inductance) in each motor, so the fault isolation is achieved.

6.4 Simulation

6.4.1 Parameter setting and thresholds computation

A two-link manipulator driven by DC motors as in Fig 6.1 is considered, the system model is as (6.14) with the following parameters ([91]):

$$M(q) = \begin{bmatrix} 7.5 + 2.5\cos(q_2) & 1.25 + 2.5\cos(q_2) \\ 1.25 + 2.5\cos(q_2) & 1.25 \end{bmatrix},$$

$$C(q, \dot{q}) = \begin{bmatrix} -2.5\sin(q_2)\dot{q}_2 & -2.5\sin(q_2)(\dot{q}_1 + \dot{q}_2) \\ 2.5\sin(q_2)\dot{q}_1 & 0 \end{bmatrix},$$

$$g(q) = \begin{bmatrix} 98\cos(q_1) + 24.5\cos(q_1 + q_2) \\ 24.5\cos(q_1 + q_2) \end{bmatrix},$$

$$K = \begin{bmatrix} 6 & 0 \\ 0 & 5 \end{bmatrix}, \quad R_e = \begin{bmatrix} 1.2 & 0 \\ 0 & 1.2 \end{bmatrix},$$

$$Q_{in} = \begin{bmatrix} 0.05 & 0 \\ 0 & 0.05 \end{bmatrix}, \quad W_t = \begin{bmatrix} 0.8 & 0 \\ 0 & 0.8 \end{bmatrix}$$

where $q = \begin{bmatrix} q_1 & q_2 \end{bmatrix}^T$. The bounds of unknown inputs defined in (6.8) and (6.13) are

$$\delta_m = 0.47, \quad \delta_c = 0.28, \quad \delta_{c1} = 0.2, \quad \delta_{c2} = 0.2,$$

and the armature voltage (input) is set as:

$$u_v = \begin{bmatrix} 60(\sin(2t) + \cos(2t)) & 40(\sin(2t) + \cos(2t)) \end{bmatrix}^T$$

Driven by the above input signal, the bounds of the system states defined in (6.32) are

$$\delta_{\dot{q}} = 10.82, \quad \delta_i = 85, \quad \delta_{i1} = 72.10, \quad \delta_{i2} = 46.57.$$

The width of the evaluation moving time window and the isolation window are set as $\eta = 0.1s$ and $w = 0.3s$. Then according to (6.34), (6.38) and (6.45), we have the thresholds as

$$\Theta = 2.91, \quad \Theta_M = 0.51, \quad \Theta_D = 2.40, \quad \Theta_{D1} = 1.44, \quad \Theta_{D2} = 0.93.$$

6.4.2 Simulation results

For different kinds of faults, simulations of 400 seconds have been carried out. In all simulations, the faults appear at $t = 200$ seconds. In figures, the solid line represents the behavior of $\mathcal{H}(\tau)$ or $\mathcal{R}_i(\tau)(i = 1, 2)$, and the dashed line represents the thresholds Θ and $-\Theta$.

6.4 Simulation

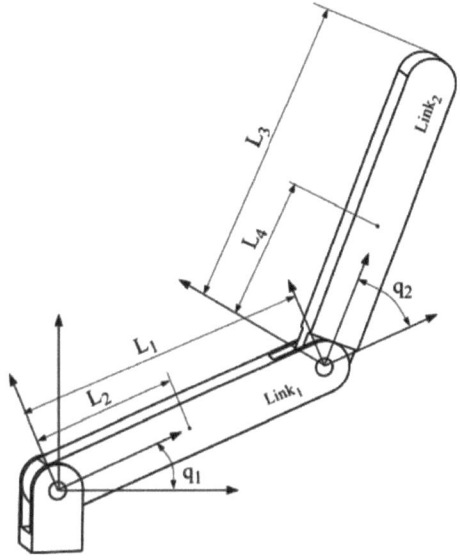

Figure 6.1: Two-link manipulator driven by DC motors

Friction rate change of manipulator. Fig 6.2 to Fig 6.5 show the simulation results, when friction rate is faulty. Two abrupt faults are considered. For the first fault which leads to the change of the friction rate of the second link $\Delta K_2 = 3$, the behavior of $\mathscr{H}(\tau)$ is as Fig 6.2a. We can see that, after the fault appears, it turns to be $|\mathscr{H}(\tau)| > \Theta$, which leads to a successful fault detection. $\mathscr{H}_M(\tau)$ and $\mathscr{H}_D(\tau)$ for subsystems are as in Fig 6.2b and Fig 6.2c, since only $\mathscr{H}_M(\tau)$ crosses the threshold after the occurance of the fault, the fault in manipulator subsystem is successfully isolated. On the other hand, as $\mathscr{H}_M(\tau)$ only crosses one of the thresholds, according to the isolation logic (6.40) dissipated-energy change (friction rate change) is isolated. In order to further isolate in which link friction rate is changed, the behavior of $\mathscr{R}_{M1}(\tau)$ and $\mathscr{R}_{M2}(\tau)$ are shown in Fig 6.3a and Fig 6.3b. We can see that, for $\tau > \tau_{det} + w$ ($\tau_{det} = 209.1s, w = 0.3s$), only $\mathscr{R}_{M2}(\tau)$ stays under the thresholds ($\mathscr{R}_{M2}(\tau)$ crosses the threshold only when $\tau < \tau_{det} + w$). Since $\mathscr{H}_M(\tau) \leq \Theta$ is fulfilled when $|M_{Q_2}(\tau)| \leq 0.001$ as shown in Fig 6.3c, according to isolation logic (6.44), the fault which leads to friction change in the second link is successfully isolated. For the second fault which leads to the change of the friction rate of the first link $\Delta K_1 = 6$, the behavior of $\mathscr{H}(\tau)$, $\mathscr{H}_M(\tau)$ and $\mathscr{H}_D(\tau)$ is as Fig

6.4, similarly friction rate change in manipulator subsystem is isolated. $\mathcal{R}_{M1}(\tau)$, $\mathcal{R}_{M2}(\tau)$ and $\mathcal{H}_M(\tau)$ (when $|M_{Q_1}(\tau)| \leq 0.001$) are shown in Fig 6.5, according to (6.44), friction rate change in the first link is successfully isolated.

Unexpected load in manipulator. Fig 6.6 shows the results when there is an unexpected load $m = 3kg$ in the second link of manipulator. From the behavior of $\mathcal{H}(\tau)$ we can see that, the fault has been successfully detected after it appears at $t = 200$ seconds. And it can be observed from the behavior of $\mathcal{H}_M(\tau)$ and $\mathcal{H}_D(\tau)$ that, fault in manipulator subsystem is isolated. Since $\mathcal{H}_M(\tau)$ crosses both thresholds, according to the isolation logic (6.40), stored-energy change in manipulator subsystem is successfully isolated.

Resistance change in DC motors. Fig 6.7 to Fig 6.8 show the results when there is an abrupt fault which leads to a resistance change $\Delta R_{e2} = 0.3\Omega$ in the second DC motor. It can be observed from Fig 6.7 that, the fault has been successfully detected by $\mathcal{H}(\tau)$ and isolated as a fault in DC motor subsystem by $\mathcal{H}_M(\tau)$ and $\mathcal{H}_D(\tau)$. For the isolation between two DC motors, $\mathcal{H}_{D1}(\tau)$ and $\mathcal{H}_{D2}(\tau)$ are shown in Fig 6.8, we can see that, only $\mathcal{H}_{D2}(\tau)$ crosses the threshold, so fault in the second DC motor is isolated. Since $\mathcal{H}_{D2}(\tau)$ only crosses one of the threshold, according to isolation logic (6.48), finally dissipated-energy change (resistance change) in the second DC motor is successfully isolated.

Inductance change in DC motors. Fig 6.9 to Fig 6.10 show the results when there is an abrupt fault which leads to an inductance change $\Delta Q_{i1} = -0.02H$ in the first DC motor. It can be observed from Fig 6.9 that, the fault is successfully detected by $\mathcal{H}(\tau)$ and also isolated as a fault in DC motor subsystem by $\mathcal{H}_M(\tau)$ and $\mathcal{H}_D(\tau)$. For the isolation between two DC motors, $\mathcal{H}_{D1}(\tau)$ and $\mathcal{H}_{D2}(\tau)$ are shown in Fig 6.10, we can see that, only $\mathcal{H}_{D1}(\tau)$ crosses the threshold, so fault in the first DC motor is isolated. Since $\mathcal{H}_{D1}(\tau)$ crosses both thresholds, according to isolation logic (6.48), stored-energy change (inductance change) in the first DC motor is successfully isolated.

6.5 Summary

In this chapter, the use of proposed energy-balance based FDI framework has been demonstrated by robot manipulator benchmark study. The stored, dissipated and supplied energies of the benchmark system have been studied, and energy balances for the whole system as well as for two subsystems have been established. Based on them, energy-balance based fault detection and isolation schemes were designed. The complete FDI system together with the benchmark system was implemented in MATLAB/SIMULINK. The results show that, the proposed FDI schemes can effectively detect and isolate the faults in robot manipulator benchmark.

Chapter 6 Application to the robot manipulator benchmark

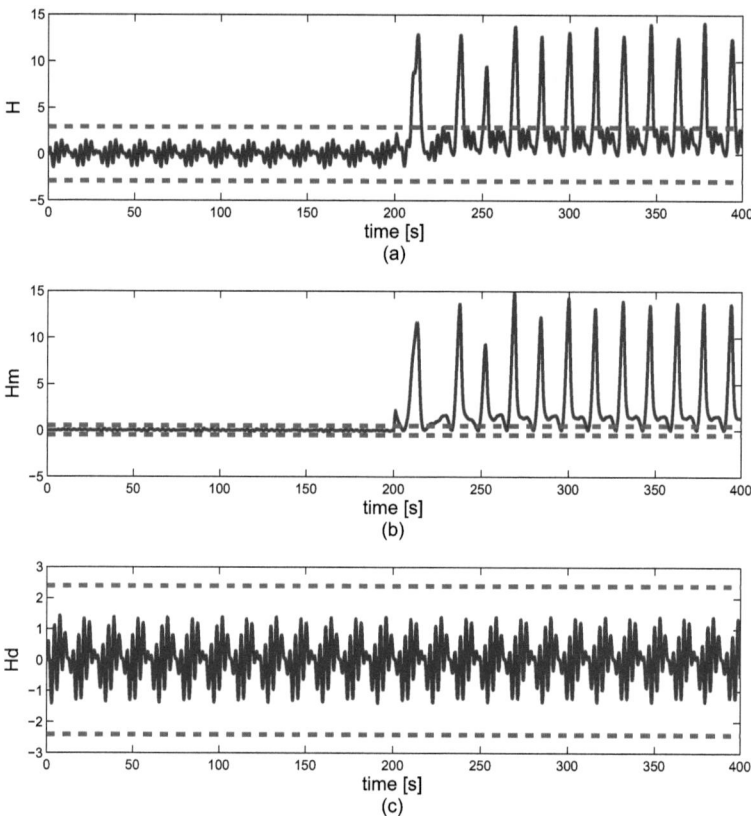

Figure 6.2: Friction rate change in the second link of manipulator. Figure (a), (b) and (c) show the behavior of $\mathscr{H}(\tau)$, $\mathscr{H}_M(\tau)$ and $\mathscr{H}_D(\tau)$, when when an abrupt friction rate change $\Delta K_2 = 3$ of the second link occurs at $t = 200$ seconds.

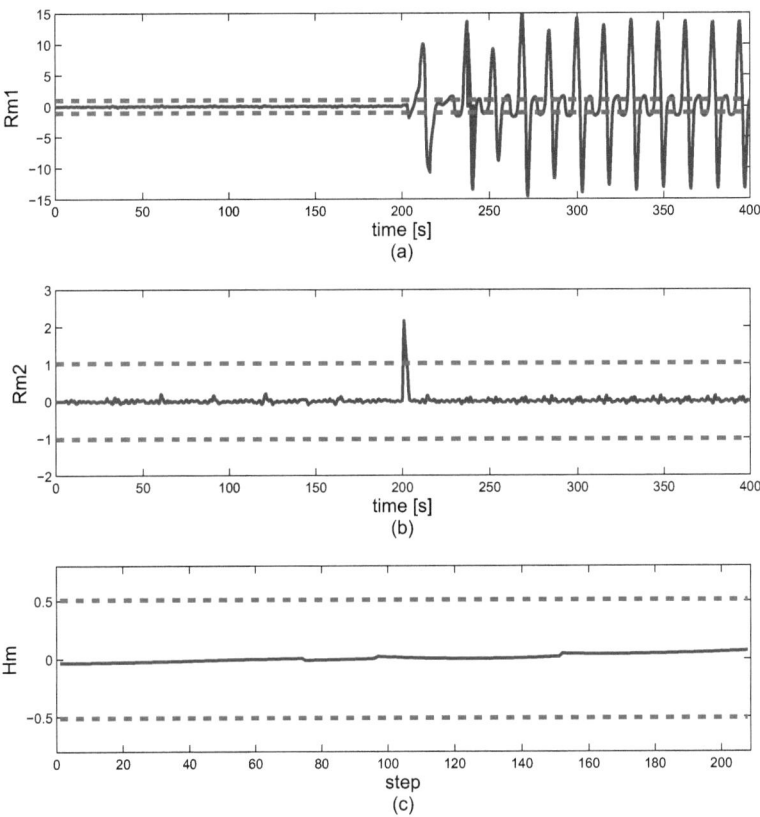

Figure 6.3: Friction rate change in the second link of manipulator. Figure (a) and (b) show the behavior of $\mathscr{R}_{M1}(\tau)$ and $\mathscr{R}_{M2}(\tau)$ for the fault $\Delta K_2 = 3$ at $t = 200$ seconds. Figure (c) shows the behavior of $\mathscr{H}_M(\tau)$ when $|M_{Q_2}(\tau)| \leq 0.001$.

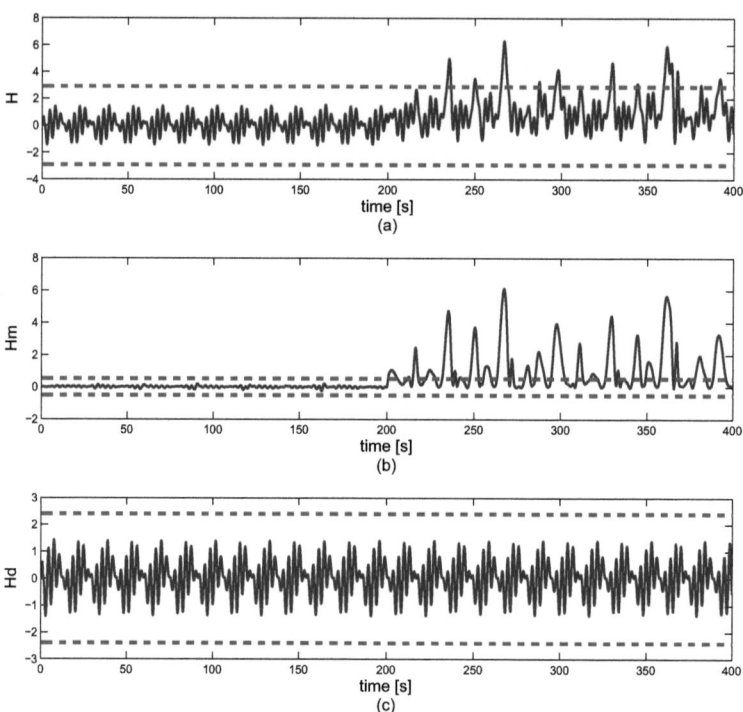

Figure 6.4: Friction rate change in the first link of manipulator. Figure (a), (b) and (c) show the behavior of $\mathscr{H}(\tau)$, $\mathscr{H}_M(\tau)$ and $\mathscr{H}_D(\tau)$, when when an abrupt friction rate change $\Delta K_1 = 6$ of the first link occurs at $t = 200$ seconds.

6.5 Summary

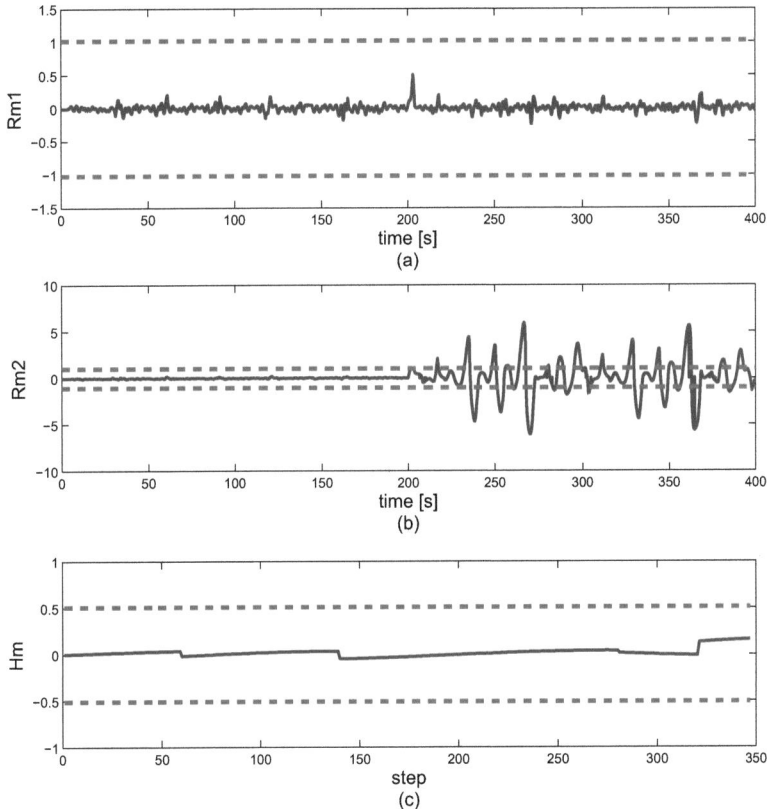

Figure 6.5: Friction rate change in the first link of manipulator. Figure (a) and (b) show the behavior of $\mathscr{R}_{M1}(\tau)$ and $\mathscr{R}_{M2}(\tau)$ for the fault $\Delta K_1 = 6$ at $t = 200$ seconds. Figure (c) shows the behavior of $\mathscr{H}_M(\tau)$ when $|M_{Q_1}(\tau)| \leq 0.001$.

Chapter 6 Application to the robot manipulator benchmark

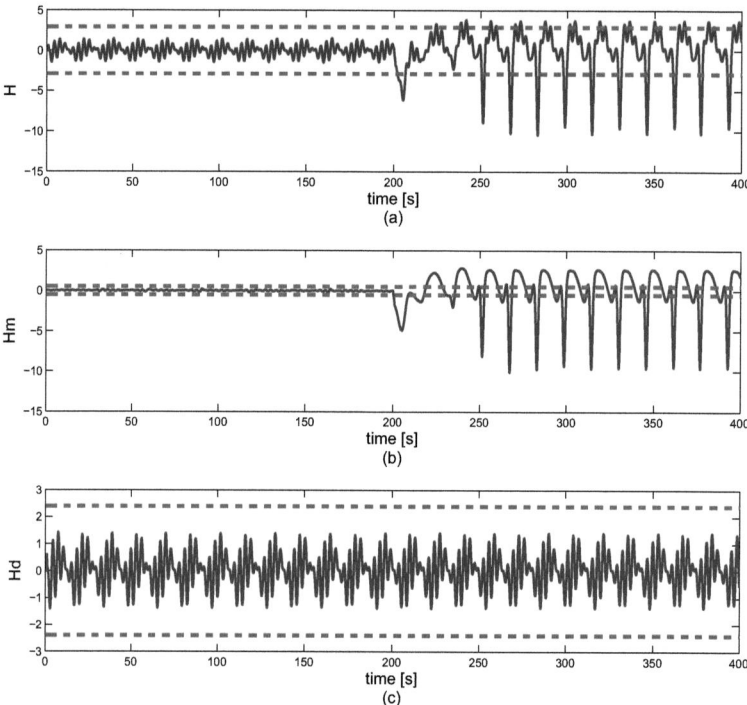

Figure 6.6: Unexpected load in manipulator. Figure (a), (b) and (c) show the behavior of $\mathscr{H}(\tau)$, $\mathscr{H}_M(\tau)$ and $\mathscr{H}_D(\tau)$, when there is an unexpected load $m = 3kg$ appears in the second link at $t = 200$ seconds.

6.5 Summary

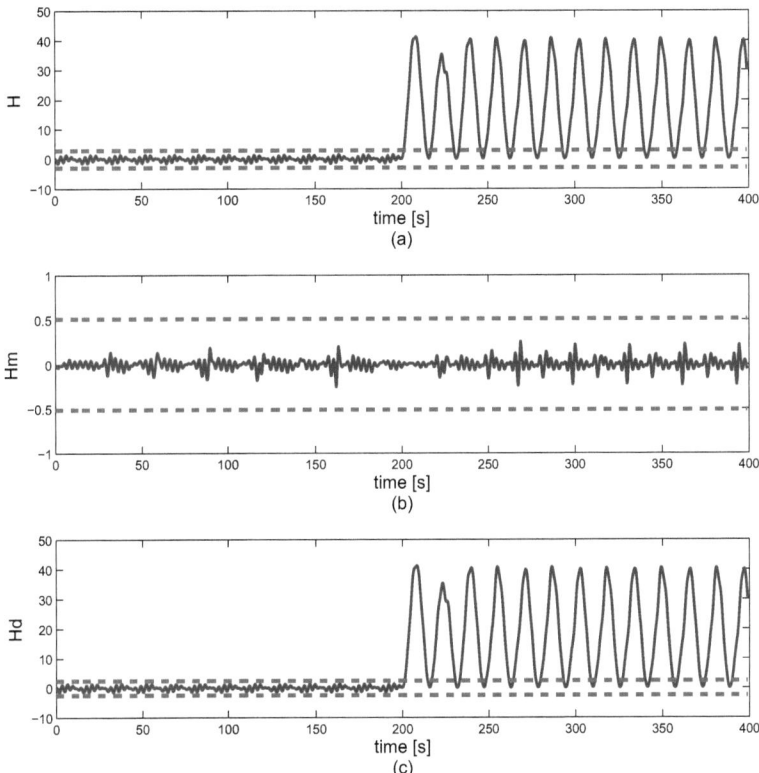

Figure 6.7: Resistance change in the second DC motor. Figure (a), (b) and (c) show the behavior of $\mathcal{H}(\tau)$, $\mathcal{H}_M(\tau)$ and $\mathcal{H}_D(\tau)$, when an abrupt resistance change $\Delta R_{e2} = 0.3\Omega$ of the second DC motor occurs at $t = 200$ seconds.

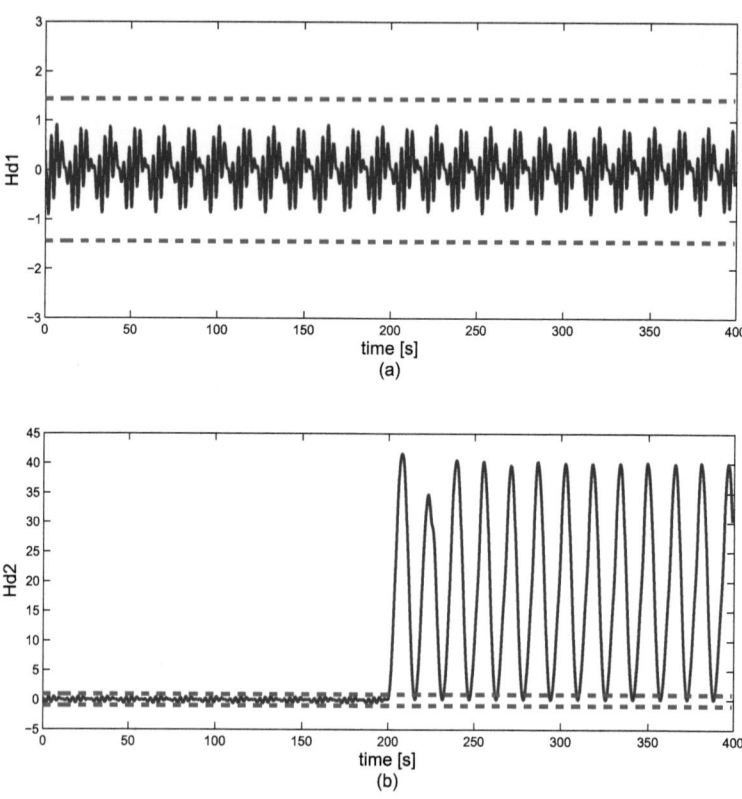

Figure 6.8: Resistance change in the second DC motor. Figure (a) and (b) show the behavior of $\mathcal{H}_{D1}(\tau)$ and $\mathcal{H}_{D2}(\tau)$ for the fault $\Delta R_{e2} = 0.3\Omega$ at $t = 200$ seconds.

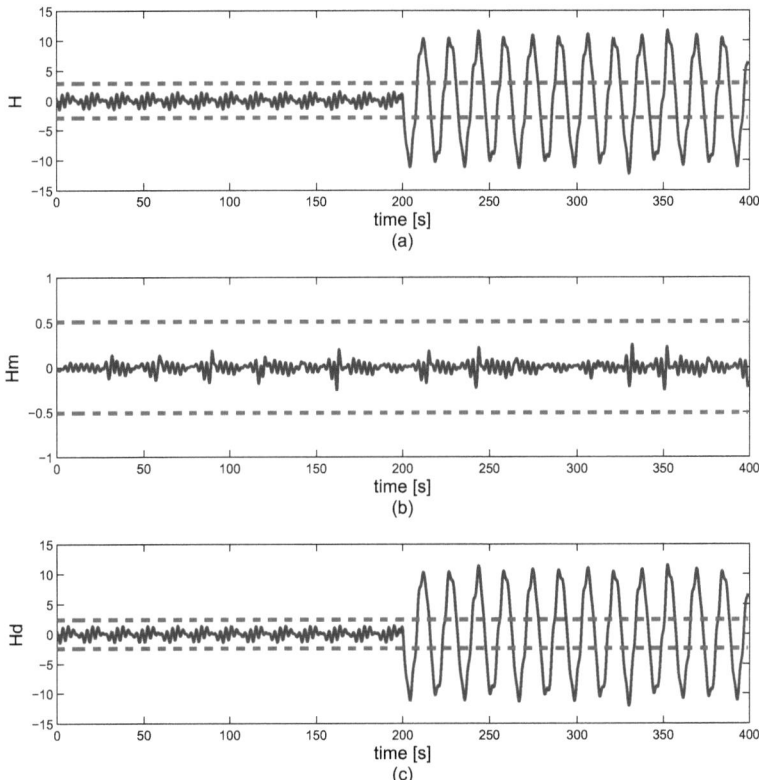

Figure 6.9: Inductance change in the first DC motor. Figure (a), (b) and (c) show the behavior of $\mathcal{H}(\tau)$, $\mathcal{H}_M(\tau)$ and $\mathcal{H}_D(\tau)$, when an abrupt inductance change $\Delta Q_{i1} = -0.02H$ of the first DC motor occurs at $t = 200$ seconds.

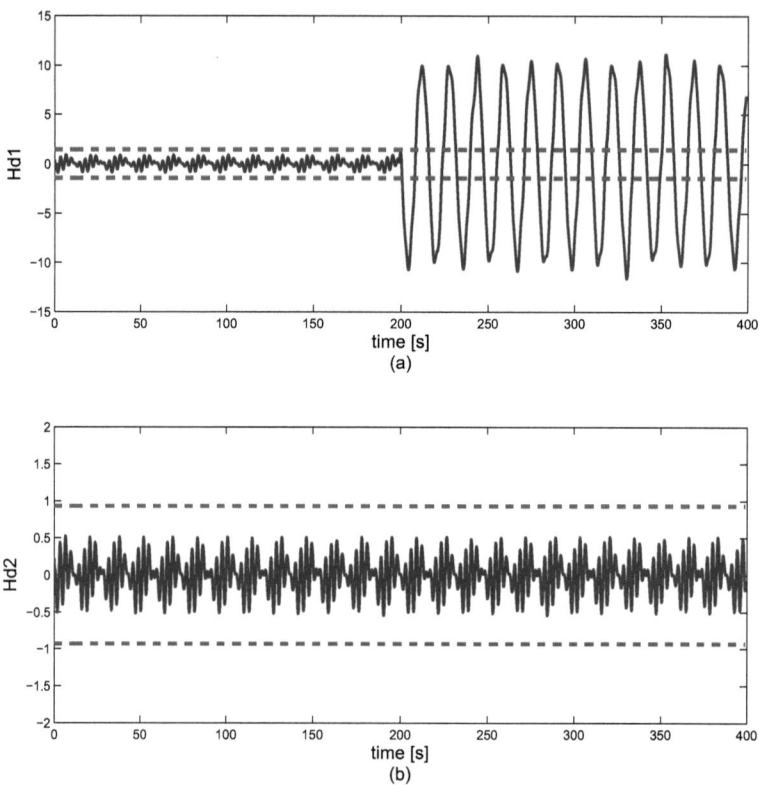

Figure 6.10: Inductance change in the first DC motor. Figure (a) and (b) show the behavior of $\mathcal{H}_{D1}(\tau)$ and $\mathcal{H}_{D2}(\tau)$ for the fault $\Delta Q_{i1} = -0.02H$ at $t = 200$ seconds.

Chapter 7
Conclusions and future directions

This chapter summarizes the results obtained in this dissertation, and presents the concluding remarks. Some directions for further developments of the proposed approaches are also suggested.

7.1 Conclusions

The major focus of this thesis is on fault detection and isolation for nonlinear systems. Two kinds of nonlinear systems are considered, which are Lipschitz nonlinear systems and passive nonlinear systems. The performance of the proposed FDI schemes have been illustrated by academical examples and benchmark studies.

The first objective of this thesis is to design an FD system for Lipschitz nonlinear process which results in a trade-off between low false alarm rate and high fault detection rate in the norm-based framework. To achieve this objective, residual generation and evaluation have been desgined in an integrated way, and the optimization problem was formulated as: given FDR, minimizing the FAR. An iterative algorithm based on linear matrix inequalities is proposed to solve the optimization problem. A desgin example was provided to illustrate the proposed methodologies. From the discussions presented in Chapter 3, it is concluded that:

A better performance of the FD system for Lipschitz nonlinear processes can be achieved by designing the residual generation and evaluation in an integrated way.

The practical requirements of FDR and FAR can be used as indices in the integrated design approach, which makes the application of the FD system much easier.

The integrated design of FD system for Lipschitz nonlinear processes can be solved iteratively based on linear matrix inequalities.

Chapter 7 Conclusions and future directions
___

The second objective is to eatablish an energy-balance based FDI framework for passive nonlinear systems. Passivity is an important system property which relates closely to system energies. Based on it, energy balance for passive system has been constructed and used for fault detection. Faults were defined based on their influences to system energies, and a two-step fault isolation scheme was proposed. The first step of the fault isolation is to find out which kind of fault (energy change) appears in the system, and the second step is to isolate the faulty component. The desgin proceedure of the proposed energy-based FDI framework for input-affine nonlinear systems and Lagrangian nonlinear systems have been studied in Chapter 5. The application to robot manipulator benchmark shows the effectiveness of the proposed energy-based FDI schemes. From the discussions presented in these chapters, the following conclusions can be drawn:

> The fault detection and isoaltion of passive nonlinear systems can be achieved in an energy-balance based framework, which has a simple design proceedure and requires little on-line computation.
>
> An energy balance which contains stored, dissipated and supplied energies can be constructed for passive nonlinear systems.
>
> Fault detection can be achieved by checking the validity of the energy balance.
>
> Faults can be defined according to their influences on system energies. Based on this defintion, faults in energy-dissipating components and energy-storing components can be distinguished easily.
>
> Fault isolation can be carried out in a two-step form in the energy-balance based framework, which can considerably reduce the complexity of the isolation scheme.
>
> For input-affine nonlinear systems and Lagrangian nonlinear systems, the energy-balance based FDI schemes can be designed effectively.
>
> For complex systems, energy balances of subsystems can be used for fault isolation.

7.2 Future directions

The preceding section summarized the results obtained in this dissertation. The proposed techniques and their application to improve the performance of FDI system of nonlinear processes were briefly described. Besides the admired features of the proposed methods, there is a

7.2 Future directions

room for further improvements. In the following, some possible research directions for further extension of the proposed FDI schemes are outlined.

Chapter 3 proposed the integrated design approach of observer-based FD for Lipschitz nonlinear systems, which results in a trade-off between low false alarm rate and high fault detection rate in the norm-based framework. Since there are model uncertainties, system input u will also influence the residual signal which could lead to false alarms. In the proposed approach, input u has been summed up with disturbance d into one vector $d_0 = \begin{bmatrix} u \\ d \end{bmatrix}$, and the influence of d_0 to the residual signal is minimized to achieve a lower false alarm rate. In this way, u has been treated as an unknown input, which could lead to conservative results. Since u is online available, a possible future direction to extend the proposed approach is to make use of the information of input u in the FD system design, for instance, adaptive threshold etc. Another future direction is to extend the integrated FD system design approach to more general nonlinear processes.

Chapter 4 proposed the energy-balance based FDI framework for passive nonlinear systems, and the design proceedure of the proposed FDI framework has been studied for two classes of nonlinear systems in Chapter 5 and 6. In the proposed FDI schemes, the construction of the energy balance is based on an input-output system model, which may not be available for some complex processes like in chemical industry. In [19], assume the process is in the steady-state, a so-called signals energy balance is constructed by identifying the parameters using the fault-free data. A possible future direction is to extend the proposed FDI schemes based on the identified energy balance in [19]. In [20], besides energy balance, the mass balance of the process has also been established and used for fault detection purpose. Since for many chemical processes, mass balance is a very important property, another future direction is to extend the proposed energy-balance based FDI framework to a more general form which could also include the mass balance.

Bibliography

[1] I. Izadi, S. L. Shah, D. S. Shook, and T. Chen, "An Introduction to Alarm Analysis and Design," in *Proc. of IFAC SafeProcess'09*, (Barcelona, Spain), June 30-July 3, 2009.

[2] J. M. Maciejowski and C. N. Jones, "MPC fault-tolerant flight control dase study: flight 1862," in *Proc. of IFAC SafeProcess'09*, (Barcelona, Spain), June 30-July 3, 2009.

[3] R. J. Patton, "Fault-tolerant control: the 1997 situation," in *Proc. of IFAC SafeProcess'09*, (Kingston Upon Hull, UK), 1997.

[4] M. M. Mahmoud, J. Jiang, and Y. Zhang, *Active fault tolerant control systems: stochastic analysis and synthesis*. Lecture Notes in Control and Information Sciences. Springer, 2003.

[5] S. Kanev, *Robust fault-tolerant control*. PhD thesis, University of Twente, the Netherlands, 2004.

[6] S. X. Ding, *Model-based Fault Diagnosis Techniques*. Springer, 2008.

[7] J. Chen and R. J. Patton, *Robust model-based fault diagnosis for dynamic systems*. Boston: Kluwer Academic Publishers, 1999.

[8] K. Pattan, *Artificial Neural Networks for the Modelling and Fault Diagnosis of Technical Processes*. Springer, 2008.

[9] J. J. Gertler, *Fault Detection and Diagnosis in Engineering Systems*. Marcel Dekker, 1998.

[10] R. Isermann, *Fault-Diagnosis Systems*. Springer, 1988.

[11] M. Blanke, M. Kinnaert, J. Lunze, and M. Staroswiecki, *Diagnosis and Fault Tolerant Control*. Springer, 2nd ed., 2006.

[12] P. M. Frank, "On-line fault detection in uncertain nonlinear systems using diagnostic observers: a survey," *International Journal of Systems Science*, vol. 25, no. 12, pp. 2129–2154, 1994.

[13] E. A. Garcia and P. M. Frank, "Deterministic Nonlinear observer-based approaches to fault diagnosis: a survey," *Control Eng. Practice*, vol. 5, pp. 663–670, 1997.

[14] M. Witczak, *Identification and fault detection of non-linear dynamic systems*. Lecture Notes in Control and Computer Science. University of Zielona, Gora Press, 2003.

[15] R. Rajamani, "Observers for Lipschitz Nonlinear Systems," *IEEE Transactions on Automatic Control*, vol. 43, no. 3, pp. 397–401, 1998.

[16] A. M. Pertew, H. J. Marquez, and Q. Zhao, "LMI-based sensor fault diagnosis for nonlinear Lipschitz systems," *Automatica*, vol. 43, pp. 1464–1469, 2007.

[17] W. Chen and M. Saif, "Observer-based strategies for actuator fault detection, isolation and estimation for certain class of uncertain nonlinear systems," *IET Control Theory & Applications*, vol. 1, no. 6, pp. 1672–1680, 2007.

[18] R. Rajamani and A. Ganguli, "Sensor fault diagnostics for a class of non-linear systems using linear matrix inequalities," *International Journal of Control*, vol. 77, no. 10, pp. 920–930, 2004.

[19] D. Theilliol, H. Noura, D. Sauter, and F. Hamelin, "Sensor fault diagnosis based on energy balance evaluation: Application to a metal processing," *ISA Transactions*, vol. 45, no. 4, pp. 603–610, 2006.

[20] A. Berton and D. Hodouin, "Linear and bilinear fault detection and diagnosis based on mass and energy balance equations," *Control Engineering Practice*, vol. 11, pp. 103–113, 2003.

[21] H. Yang, V. Cocquempot, and B. Jiang, "Fault Tolerance Analysis for Switched Systems Via Global Passivity," *IEEE Trans. Circuits and systems*, vol. 55, no. 12, pp. 1279–1283, 2008.

[22] H. K. Khalil, *Nonlinear Systems*. Prentice Hall, 1996.

[23] P. Zhang and S. X. Ding, "An integrated trade-off design of observer based fault detection systems," *Automatica*, vol. 44, pp. 1886–1894, 2008.

[24] S. X. Ding, P. M. Frank, E. L. Ding, and T. Jeinsch, "Fault detection system design based on a new trade-off strategy," in *Proceedings of the 39th IEEE Conference on Decision and Control*, (Sydney, Australia), December 2000.

Bibliography

[25] R. Isermann and P. Balle, "Trends in the application of model-based fault detection and diagnosis of technical processes," *Control Eng. Practice*, vol. 5, pp. 709–719, 1997.

[26] P. M. Frank, "Analytical and qualitative model-based fault diagnosis-a survey," *Europ. J. Control*, vol. 2, pp. 6–28, 1996.

[27] J. Wuennenberg and P. M. Frank, "sensor fault detection via robust observers," in *First European Workshop On Fault Diagnostics. Reliability and Related Knowledge-Based Approaches*, (Rhodes, Greece), 1986.

[28] W. Ge and C. Z. Fang, "Extended robust observation approach for failure isolation," *Int. J. Control*, vol. 49:5, pp. 1537–1553, 1989.

[29] P. M. Frank and J. Wuennenberg, *Fault Diagnosis in Dynamic Systems: Theory and Application*, ch. Robust fault diagnosis using unknown input schemes, pp. 47–98. Prentice Hall, 1989.

[30] P. M. Frank and X. Ding, "Frequency domain approach to optimally robust residual generation and evaluation for model-based fault diagnosis," *Automatica*, vol. 30, pp. 789–804, 1994.

[31] S. X. Ding, L. Guo, and P. M. Frank, "A frequency domain approach to fault detection of uncertain dynamic systems," in *Proceedings of the 32th IEEE Conference on Decision and Control*, pp. 1722–1727, 1993.

[32] Z. Qiu and J. Gertler, "Robust FDI systems and h-infinity optimization," in *Proceedings of the American Control Conference*, pp. 1710–1715, 1993.

[33] M. Hou and R. J. Patton, "An LMI approach to infinity fault detection observers," in *Proceedings of the UKACC international conference on control*, pp. 1710–1715, 1996.

[34] J. L. Wang, G. Yang, and J. Liu, "An LMI approach to H_- index and mixed H_-/H_∞ fault detection observer design," *Automatica*, vol. 43, pp. 1656–1665, 2007.

[35] D. Henry and A. Zolghadri, "Design and analysis of robust residual generators for systems under feedback control," *Automatica*, vol. 41, no. 2, pp. 251–264, 2005.

[36] S. X. Ding, P. M. Frank, E. L. Ding, and T. Jeinsch, "A unified approach to the optimization of fault detection systems," *International Journal of Adaptive Control and Signal Process*, vol. 14, pp. 725–745, 2000.

[37] P. Zhang, S. X. Ding, G. Wang, and D. Zhou, "Fault detection of linear discrete-time periodic systems," *IEEE Transactions on Automatic Control*, vol. 50, no. 2, pp. 239–244, 2005.

[38] E. Y. Chow and A. S. Willsky, "Analytical redundancy and the design of robust failure detection systems," *IEEE Trans. Automat. Contr.*, vol. 29, pp. 603–614, 1984.

[39] J. Gertler and D. Singer, "A new structural framework for parity equation based failure detection and isolation," *Automatica*, vol. 26, pp. 381–388, 1990.

[40] J. Gertler and M. Kunert, "Optimal residual decoupling for robust fault diagnosis," *Int. J. Control*, vol. 61, pp. 395–421, 1995.

[41] G. Delmaire, J. P. Cassar, and M. Staroswiecki, "Comparision of identification and parity space approach for failure detection in single-input single-output systems," in *Proc. IEEE Conf. Contr. App.*, (Glasgow, UK), pp. 865–870, 1994.

[42] M. Basseville and I. V. Nikiforov, *Detection of abrupt changes-theory and application*. Prentice-Hall, 1993.

[43] M. Kinnaert, "Model-based statistical signal processing and decision theoretical approach to monitoring," in *Proc. IFAC SAFEPROCESS*, (Washington, USA), pp. 1–12, 2003.

[44] T. L. Lai and J. Z. Shan, "Efficient recursive algorithms for detection of abrupt changes in signals and control systems," *IEEE Transactions on Automatic Control*, vol. 44, pp. 952–966, 1999.

[45] J. P. Gauthier, H. Hammouri, and S. Othman, "A simple observer for nonlinear systems application to biorectors," *IEEE Trans. Automat. Contr.*, vol. 37(6), pp. 875–880, 1992.

[46] H. Hammouri, M. Kinnaert, and E. H. E. Yaagoubi, "Observer-based approach to fault detection and isolation for nonlinear systems," *IEEE Trans. Automat. Contr.*, vol. 44(10), pp. 1879–1884, 1999.

[47] R. M. Guerra, R. Garrido, and A. O. Miron, "The fault detection problem in nonlinear systems using residual generators," *IMA J. Math. Contr. Infor.*, vol. 22, pp. 119–136, 2005.

[48] G. Besancon, "High-gain observation with disturbance attenuation and application to fault detection and isolation for nonlinear systems," *Automatica*, vol. 39, pp. 1095–1102, 2003.

[49] X. G. Yan and C. Edwards, "Sensor fault detection and isolation for nonlinear systems based on sliding mode observer," vol. 21, pp. 657–673, 2007.

[50] X. G. Yan and C. Edwards, "Nonlinear robust fault reconstructive and estimation using a sliding mode observers," vol. 43, pp. 1605–1614, 2007.

[51] X. G. Yan and C. Edwards, "Robust sliding mode observer-based actuator fault detection and isolation for a class for nonlinear systems," vol. 39, pp. 1605–1614, 2008.

[52] M. A. Massoumia, "A geometric approach to the synthesis of failure detection filters," *IEEE Trans. Automat. Contr.*, vol. 31, pp. 839–845, September 1986.

[53] C. D. Persis and A. Isidori, "A geometric approach to nonlinear fault detection and isolation," *IEEE Trans. Automat. Contr*, vol. 46, pp. 853–865, 2001.

[54] H. Hammouri, P. Kabore, and M. Kinnaert, "A Geometric Approach to Fault detection and Isolation for Bilinear Systems," *IEEE Trans. Automat. Contr.*, vol. 46(9), pp. 1451–1455, 2001.

[55] C. D. Persis and A. Isidori, *Nonlinear Control in the Year 2000*, ch. An H_∞-optimal Fault Detection Filter for Bilinear Systems, pp. 331–3393. Springer-Verlag, 2000.

[56] E. Yaz and A. Azemi, "Actuator fault detection and isolation in nonlinear systems using LMIs and LMEs," in *Proceedings of the American Control Conference*, (Philadelphia, Pennsylvania), June 1998.

[57] C. E. de Souza, L. Xie, and Y. Wang, "H_∞ filtering for a class of uncertain nonlinear systems," *Systems and Control Letters*, vol. 20, no. 6, p. 419426, 1993.

[58] L. Xie, C. E. de Souza, and Y. Wang, "Robust filtering for a class of discrete-time uncertain nonlinear systems: An H_∞ approach," *International Journal of Robust and Nonlinear Control*, vol. 6, no. 4, pp. 297–312, 1996.

[59] M. Abbaszadeh and H. Marquez, "LMI Optimization Approach to Robust H_∞ filtering for discrete-time nonlinear uncertain systems," in *Proceedings of the American Control Conference*, (Washington, USA), June 2008.

[60] M. Blanke, M. Kinnaert, J. Lunze, and M. Staroswiecki, *Diagnosis and fault-tolerant control*. Springer, 2003.

[61] R. J. Patton, P. M. Frank, and R. N. Clark, *Issues of fault diagnosis for dynamic systems*. Springer, 2002.

[62] H. Hammouri, M. Kinnaert, and E. H. E. Yaagoubi, "Observer-Based Approach to Fault Detection and Isolation for Nonlinear systems," *IEEE Transactions on Automatic Control*, vol. 44, no. 10, pp. 1879–1884, 1999.

[63] R. Ferrari, T. Parisini, and M. Polycarpou, "A Fault Detection and Isolation Scheme for Nonlinear Uncertain Discrete-Time systems," in *Proceedings of the 46th IEEE Conference on Decision and Control*, (New Orleans, LA, USA), December 2007.

[64] S. Narasimhan, P. Vachhani, and R. Rengaswamy, "New nonlinear residual feedback observer for fault diagnosis in nonlinear systems," *Automatica*, vol. 44, no. 9, pp. 2222–2229, 2007.

[65] A. Shumsky, "Redundancy relations for fault diagnosis in nonlinear uncertain systems," *International Journal of Applied Mathematics and Computer Science*, vol. 17, no. 4, pp. 477–489, 2007.

[66] A. Edelmayer, J. Bokor, Z. Szab, and F. Szigeti, "Input reconstruction by means of system inversion: A geometric approach to fault detection and isolation in nonlinear systems," *International Journal of Applied Mathematics and Computer Science*, vol. 14, no. 2, pp. 189–199, 2004.

[67] A. Q. Khan, M. Abid, W. Chen, and S. X. Ding, "On optimal Fault Detection of nonlinear systems," in *Proceedings of the 48th IEEE Conference on Decision and Control*, (Shanghai, China), pp. 1032–1037, December 2009.

[68] Y. Wang, L. Xie, and C. E. Souza, "Robust control of a class of uncertain nonlinear systems," *Systmes & Control Letters*, vol. 19, pp. 139–149, 1992.

[69] R. Ortega, A. J. V. D. Schaft, I. Mareels, and B. Maschke, "Putting energy back in control," *IEEE Transactions on Automatic Control*, vol. 21, no. 2, pp. 18–33, 2001.

[70] J. C. Willems, "Dissipative dynamical systems," *Eur.J.Contr.*, vol. 13, pp. 134–151, 2007.

[71] C. I. Byrnes, A. Isidori, and J. C. Willems, "Passivity, Feedback Equivalence, and the Global Stabilization of Minimum Phase Nonlinear Systems," *IEEE Trans. Automat. Contr*, vol. 36, no. 11, pp. 1228–1240, 1991.

Bibliography

[72] A. J. Koshkouei, "Passivity-based sliding mode control for nonlinear systems," *International Journal of adaptive control and signal processing*, vol. 22, pp. 859–874, 2008.

[73] J. C. Travieso-Torres, M. A. Duarte-Mermoud, and D. I. Sepuleveda, "Passivity-based control for stabilization, regulation and tracking purposes of a class of nonlinear systems ," *International Journal of adaptive control and signal processing*, vol. 21, pp. 582–602, 2007.

[74] Z. P. Jiang and D. J. Hill, "Passivity and Disturbance Attenuation via Output Feedback for Uncertain Nonlinear Systems," *IEEE Trans. Automat. Contr*, vol. 43, no. 7, pp. 992–997, 1998.

[75] A. van der Schaft, $\mathcal{L}_2$-*Gain and Passivity Techniques in Nonlinear Control*. Springer, 2000.

[76] R. Lozano, B. Brogliato, O. Egeland, and B. Maschke, *Dissipative Systems Analysis and Control*. Springer, 2000.

[77] M. G. Crandall, H. Ishii, and P. L. Lions, "User's guide to viscosity solutions of second order partial differential equations," *Bulletin (New Series) of the American Mathematical Society*, vol. 27, 1992.

[78] M. Struwe, *Variational methods: applications to nonlinear partial differential equations and Hamiltonian systems*. Springer, 2008.

[79] C. T. Chen, *Linear System Theory and Design*. CBS College Publishing, 1984.

[80] P. Libermann and C. Marle, *Symplectic geometry and analytical mechanics*. Reidel, 1987.

[81] A. J. van der Schaft, *System Theoretical description of Physical Systems*. CWI Tracts 3, 1984.

[82] S. Mohseni and M. Namvar, "Fault Diagnosis in Robot Manipulators in Presence of Modeling Uncertainty and Sensor Noise," in *18th IEEE International Conference on Control Applications Part of 2009 IEEE Multi-conference on Systems and Control*, (Saint Petersburg, Russia), pp. 1750–1755, 2009.

[83] B. Reyermuth, "An Approach to Model-Based Fault Diagnosis of Industrial Robots," in *Proceedings of the IEEE International Conference on Robotics and Automation*, (Sacramento, CA), pp. 1350–1356, 1991.

[84] H. Schneider and P. M. Frank, "Observed based Supervision and Fault Detection in Robots Using Nonlinear and Fuzzy Logic Residual Evaluation," *IEEE Transactions on Control Systems Technology*, vol. 4, no. 3, pp. 274–282, 1996.

[85] F. Caccavale and I. D. Walkert, "Observer-based Fault Detection for Robot Manipulators," in *Proceedings of the 1997 IEEE International Conference on Robotics and Automation*, (Albuquerque, New Mexico), pp. 2881–2887, 1997.

[86] M. L. McIntyre, W. E. Dixon, D. M. Dawson, and I. D. Walked, "Fault Detection and Identification for Robot Manipulators," in *Proceedings of the 2004 IEEE International Conference on Robotics and Automation*, (New Orleans, LA), pp. 4981–4986, 2004.

[87] S. Nicosia and P. Tomei, "Robot Control by Using Only Measurements Joint Position," *IEEE Transactions on Automatic Control*, vol. 35, no. 9, pp. 1058–1060, 1990.

[88] J. Davila, L. Fridman, and A. Levant, "Second-Order Sliding-Mode Observer for Mechanical Systems," *IEEE Transactions on Automatic Control*, vol. 50, no. 11, pp. 1785–1789, 2005.

[89] B. Brogliato and D. Rey, "Further experimental results on nonlinear control of flexible joint manipulators," in *Proceedings of the American Control Conference*, (Philadelphia, PA, USA), pp. 2209–2211, 1998.

[90] D. C. Kay, *Schaum's outline of theory and problems of tensor calculus*. McGraw-Hill, 2006.

[91] K. W. Lee and H. K. Khalil, "Adaptive output feedback control of robot manipulators using high-gain observer," *International Journal of Control*, vol. 67, no. 6, pp. 869–886, 1997.

i want morebooks!

Buy your books fast and straightforward online - at one of world's fastest growing online book stores! Environmentally sound due to Print-on-Demand technologies.

Buy your books online at
www.get-morebooks.com

Kaufen Sie Ihre Bücher schnell und unkompliziert online – auf einer der am schnellsten wachsenden Buchhandelsplattformen weltweit! Dank Print-On-Demand umwelt- und ressourcenschonend produziert.

Bücher schneller online kaufen
www.morebooks.de

VDM Verlagsservicegesellschaft mbH
Heinrich-Böcking-Str. 6-8
D - 66121 Saarbrücken

Telefon: +49 681 3720 174
Telefax: +49 681 3720 1749

info@vdm-vsg.de
www.vdm-vsg.de

Printed by Books on Demand GmbH, Norderstedt / Germany